SAS® 9.3 Data Set Options
Reference

SAS® Documentation

The correct bibliographic citation for this manual is as follows: SAS Institute Inc. 2011. *SAS® 9.3 Data Set Options: Reference*. Cary, NC: SAS Institute Inc.

SAS® 9.3 Data Set Options: Reference

Copyright © 2011, SAS Institute Inc., Cary, NC, USA

ISBN 978-1-60764-947-2 (electronic book)
ISBN 978-1-60764-899-4

SAS Institute Inc., SAS Campus Drive, Cary, North Carolina 27513.

ISBN 978-1-60764-947-2
    1st electronic book, July 2011

ISBN 978-1-60764-899-4
    1st printing, July 2011

SAS® Publishing provides a complete selection of books and electronic products to help customers use SAS software to its fullest potential. For more information about our e-books, e-learning products, CDs, and hard-copy books, visit the SAS Publishing Web site at **support.sas.com/publishing** or call 1-800-727-3228.

Contents

About This Book

Syntax Conventions for the SAS Language

Overview of Syntax Conventions for the SAS Language

SAS uses standard conventions in the documentation of syntax for SAS language elements. These conventions enable you to easily identify the components of SAS syntax. The conventions can be divided into these parts:

- syntax components

- style conventions

- special characters

- references to SAS libraries and external files

Syntax Components

The components of the syntax for most language elements include a keyword and arguments. For some language elements only a keyword is necessary. For other language elements the keyword is followed by an equal sign (=).

keyword
> specifies the name of the SAS language element that you use when you write your program. Keyword is a literal that is usually the first word in the syntax. In a CALL routine, the first two words are keywords.

> In the following examples of SAS syntax, the keywords are the first words in the syntax:

CHAR (*string, position*)

CALL RANBIN (*seed, n, p, x*);

ALTER (*alter-password*)

BEST *w.*

REMOVE *<data-set-name>*

> In the following example, the first two words of the CALL routine are the keywords:

CALL RANBIN(*seed, n, p, x*)

> The syntax of some SAS statements consists of a single keyword without arguments:

DO;
... SAS code ...

END;

Some system options require that one of two keyword values be specified:

DUPLEX | NODUPLEX

argument
> specifies a numeric or character constant, variable, or expression. Arguments follow the keyword or an equal sign after the keyword. The arguments are used by SAS to process the language element. Arguments can be required or optional. In the syntax, optional arguments are enclosed between angle brackets.

> In the following example, *string* and *position* follow the keyword CHAR. These arguments are required arguments for the CHAR function:

> **CHAR** (*string, position*)

> Each argument has a value. In the following example of SAS code, the argument *string* has a value of 'summer', and the argument *position* has a value of 4:`x=char('summer', 4);`

> In the following example, *string* and *substring* are required arguments, while *modifiers* and *startpos* are optional.

> **FIND**(*string, substring <,modifiers> <,startpos>*

Note: In most cases, example code in SAS documentation is written in lowercase with a monospace font. You can use uppercase, lowercase, or mixed case in the code that you write.

Style Conventions

The style conventions that are used in documenting SAS syntax include uppercase bold, uppercase, and italic:

UPPERCASE BOLD
> identifies SAS keywords such as the names of functions or statements. In the following example, the keyword ERROR is written in uppercase bold:

> **ERROR**<message>;

UPPERCASE
> identifies arguments that are literals.

> In the following example of the CMPMODEL= system option, the literals include BOTH, CATALOG, and XML:

> **CMPMODEL** = BOTH | CATALOG | XML

italics
> identifies arguments or values that you supply. Items in italics represent user-supplied values that are either one of the following:

> • nonliteral arguments

>> In the following example of the LINK statement, the argument *label* is a user-supplied value and is therefore written in italics:

>> **LINK** *label*;

> • nonliteral values that are assigned to an argument

>> In the following example of the FORMAT statement, the argument DEFAULT is assigned the variable *default-format*:

FORMAT = *variable-1* <, ..., *variable-nformat*><DEFAULT = *default-format*>;

Items in italics can also be the generic name for a list of arguments from which you can choose (for example, *attribute-list*). If more than one of an item in italics can be used, the items are expressed as *item-1, ..., item-n*.

Special Characters

The syntax of SAS language elements can contain the following special characters:

=

an equal sign identifies a value for a literal in some language elements such as system options.

In the following example of the MAPS system option, the equal sign sets the value of MAPS:

MAPS = *location-of-maps*

< >

angle brackets identify optional arguments. Any argument that is not enclosed in angle brackets is required.

In the following example of the CAT function, at least one item is required:

CAT (*item-1* <, ..., *item-n*>)

|

a verticle bar indicates that you can choose one value from a group of values. Values that are separated by the vertical bar are mutually exclusive.

In the following example of the CMPMODEL= system option, you can choose only one of the arguments:

CMPMODEL = BOTH | CATALOG | XML

...

an ellipsis indicates that the argument or group of arguments following the ellipsis can be repeated. If the ellipsis and the following argument are enclosed in angle brackets, then the argument is optional.

In the following example of the CAT function, the ellipsis indicates that you can have multiple optional items:

CAT (*item-1* <, ..., *item-n*>)

'*value*' or "*value*"

indicates that an argument enclosed in single or double quotation marks must have a value that is also enclosed in single or double quotation marks.

In the following example of the FOOTNOTE statement, the argument *text* is enclosed in quotation marks:

FOOTNOTE <*n*> <*ods-format-options* '*text*' | "*text*">;

;

a semicolon indicates the end of a statement or CALL routine.

In the following example each statement ends with a semicolon: `data namegame;`
`length color name $8; color = 'black'; name = 'jack'; game =`
`trim(color) || name; run;`

References to SAS Libraries and External Files

Many SAS statements and other language elements refer to SAS libraries and external files. You can choose whether to make the reference through a logical name (a libref or fileref) or use the physical filename enclosed in quotation marks. If you use a logical name, you usually have a choice of using a SAS statement (LIBNAME or FILENAME) or the operating environment's control language to make the association. Several methods of referring to SAS libraries and external files are available, and some of these methods depend on your operating environment.

In the examples that use external files, SAS documentation uses the italicized phrase *file-specification*. In the examples that use SAS libraries, SAS documentation uses the italicized phrase *SAS-library*. Note that *SAS-library* is enclosed in quotation marks:

```
infile file-specification obs = 100;
libname libref 'SAS-library';
```

What's New in SAS 9.3 Data Set Options

Overview

The SAS data set options documentation is no longer part of *SAS Language Reference: Dictionary*. The SAS data set options that were previously documented in *SAS Language Reference: Dictionary* are now documented here, in *SAS 9.3 Data Set Options: Reference*.

Changes to *SAS Language Reference: Dictionary*

Prior to SAS 9.3, this document was part of *SAS Language Reference: Dictionary*. Starting with SAS 9.3, *SAS Language Reference: Dictionary* has been divided into seven documents:

- *SAS Data Set Options: Reference*

- *SAS Formats and Informats: Reference*

- *SAS Functions and CALL Routines: Reference*

- *SAS Statements: Reference*

- *SAS System Options: Reference*

- *SAS Component Objects: Reference* (contains the documentation for the Hash Object and the Java Object)

- *Base SAS Utilities: Reference* (contains the documentation for the SAS DATA step debugger and the SAS Utility macro %DS2CSV)

New SAS Data Set Option

The following data set option is new:

EXTENDOBSCOUNTER= (p. 19)
    extends the maximum observation count in an output SAS data file.

Recommended Reading

- *SAS Language Reference: Concepts*
- *Base SAS Glossary*
- *An Array of Challenges--Test Your SAS Skills*
- *Base SAS Procedures Guide*
- *Combining and Modifying SAS Data Sets: Examples*
- *The Little SAS Book: A Primer*
- *SAS Programming by Example*
- *Step-by-Step Programming with Base SAS Software*
- *Using the SAS Windowing Environment: A Quick Tutorial*

For a complete list of SAS publications, go to support.sas.com/bookstore. If you have questions about which titles you need, please contact a SAS Publishing Sales Representative:

SAS Publishing Sales
SAS Campus Drive
Cary, NC 27513-2414
Phone: 1-800-727-3228
Fax: 1-919-677-8166
E-mail: sasbook@sas.com
Web address: support.sas.com/bookstore

Chapter 1
Concepts

Accessibility

This document is a command-based product. For this release, no features were added to address accessibility, but the product might very well be compliant to accessibility standards because it does not have a graphical user interface, and all of its features are available to anyone who can type or otherwise produce a command. If you have specific questions about the accessibility of SAS products, send them to accessibility@sas.com or call SAS Technical Support.

Definition of Data Set Options

Data set options specify actions that apply only to the SAS data set with which they appear. They let you perform the following operations:

- renaming variables

- selecting only the first or last *n* observations for processing

- dropping variables from processing or from the output data set

- specifying a password for a data set

Syntax

Specify a data set option in parentheses after a SAS data set name. To specify several data set options, separate them with spaces.

(*option-1=value-1<...option-n=value-n>*)

These examples show data set options in SAS statements:

- `data scores(keep=team game1 game2 game3);`

- `data mydata(index=(b k) label='label for my data set' drop=p read=secret);`

- `data new(drop=i n index=(j combo=(x1 a1 a20 b1 b50 )));`

- `data idxdup2(compress=yes index=(ok1 ok2 ssn/unique ok3));`

- `proc print data=new(drop=year);`

- `set old(rename=(date=Start_Date));`

Using Data Set Options

Using Data Set Options with Input or Output SAS Data Sets

Most SAS data set options can apply to either input or output SAS data sets in DATA steps or procedure (PROC) steps. If a data set option is associated with an input data set, the action applies to the data set that is being read. If the option appears in the DATA statement or after an output data set specification in a PROC step, SAS applies the action to the output data set. In the DATA step, data set options for output data sets must appear in the DATA statement, not in any OUTPUT statements that might be present.

Some data set options, such as COMPRESS=, are meaningful only when you create a SAS data set because they set attributes that exist for the duration of the data set. To change or cancel most data set options, you must re-create the data set. You can change other options (such as PW= and LABEL=) with PROC DATASETS. For more information, see Chapter 15, "DATASETS Procedure" in *Base SAS Procedures Guide*.

When data set options appear on both input and output data sets in the same DATA or PROC step, first SAS applies data set options to input data sets. Then SAS evaluates programming statements or applies data set options to output data sets. Likewise, data set options that are specified for the data set being created are applied after programming statements are processed. For example, when using the RENAME= data set option, the new names are not associated with the variables until the DATA step ends.

In some instances, data set options conflict when they are used in the same statement. For example, you cannot specify both the DROP= and KEEP= data set options for the same variable in the same statement. Timing can also be an issue in some cases. For example, if using KEEP= and RENAME= on a data set specified in the SET statement, KEEP= needs to use the original variable names. SAS processes KEEP= before the data set is read. The new names specified in RENAME= apply to the programming statements that follow the SET statement.

How Data Set Options Interact with System Options

Many system options and data set options share the same name and have the same function. System options remain in effect for all DATA and PROC steps in a SAS job or session.

The data set option overrides the system option for the data set in the step in which it appears. In this example, the OBS= system option in the OPTIONS statement specifies that only the first 100 observations are processed from any data set within the SAS job. The OBS= data set option in the SET statement, however, overrides the system option for data set TWO and specifies that only the first five observations are read from data set TWO. The PROC PRINT step prints the data set FINAL. This data set contains the first 5 observations from data set TWO, followed by the first 100 observations from data set THREE:

```
options obs=100;

data final;
   set two(obs=5) three;
run;

proc print data=final;
run;
```

Data Set Options Documented in Other SAS Publications

In addition to data set options documented in SAS Language Reference: Dictionary, data set options are also documented in the following publications:

- *SAS Companion for Windows*

- *SAS Companion for UNIX Environments*

- *SAS Companion for z/OS*

- *SAS National Language Support: Reference Guide*

- *SAS Scalable Performance Data Engine: Reference*

- *SAS/ACCESS for Relational Databases: References*

Chapter 2
Data Set Options Dictionary

Data Set Options by Category

The categories for SAS data set options correspond to the SAS data set option groups:

Data Set Control	options that are associated with data sets
Observation Control	options that are associated with observations
User Control of SAS Index Usage	options that are associated with indexes
Variable Control	options that are associated with variables
Miscellaneous	

Category	Language elements	Description
Data Set Control	ALTER= Data Set Option (p. 8)	Assigns an ALTER password to a SAS file that prevents users from replacing or deleting the file, and enables access to a read- and write-protected file.
	BUFNO= Data Set Option (p. 9)	Specifies the number of buffers to be allocated for processing a SAS data set.
	BUFSIZE= Data Set Option (p. 11)	Specifies the size of a permanent buffer page for an output SAS data set.
	CNTLLEV= Data Set Option (p. 12)	Specifies the level of shared access to a SAS data set.
	COMPRESS= Data Set Option (p. 13)	Specifies how observations are compressed in a new output SAS data set.
	DLDMGACTION= Data Set Option (p. 15)	Specifies the action to take when a SAS data set in a SAS library is detected as damaged.
	ENCRYPT= Data Set Option (p. 17)	Specifies whether to encrypt an output SAS data set.
	EXTENDOBSCOUNTER= Data Set Option (p. 19)	Specifies whether to extend the maximum observation count in a new output SAS data file.
	GENMAX= Data Set Option (p. 22)	Requests generations for a new data set, modifies the number of generations for an existing data set, and specifies the maximum number of versions.
	GENNUM= Data Set Option (p. 23)	Specifies a particular generation of a SAS data set.
	INDEX= Data Set Option (p. 27)	Defines an index for a new output SAS data set.
	LABEL= Data Set Option (p. 31)	Specifies a label for a SAS data set.

Category	Language elements	Description
	OBSBUF= Data Set Option (p. 32)	Determines the size of the view buffer for processing a DATA step view.
	OUTREP= Data Set Option (p. 44)	Specifies the data representation for the output SAS data set.
	PW= Data Set Option (p. 47)	Assigns a READ, WRITE, and ALTER password to a SAS file, and enables access to a password-protected SAS file.
	PWREQ= Data Set Option (p. 48)	Specifies whether to display a dialog box to enter a SAS data set password.
	READ= Data Set Option (p. 48)	Assigns a READ password to a SAS file that prevents users from reading the file, unless they enter the password.
	REPEMPTY= Data Set Option (p. 51)	Specifies whether a new, empty data set can overwrite an existing SAS data set that has the same name.
	REPLACE= Data Set Option (p. 52)	Specifies whether a new SAS data set that contains data can overwrite an existing data set that has the same name.
	REUSE= Data Set Option (p. 53)	Specifies whether new observations can be written to freed space in compressed SAS data sets.
	ROLE= Data Set Option (p. 54)	Identifies the fact table for a star schema join.
	SORTEDBY= Data Set Option (p. 55)	Specifies how a data set is currently sorted.
	SPILL= Data Set Option (p. 57)	Specifies whether to create a spill file for non-sequential processing of a DATA step view.
	TOBSNO= Data Set Option (p. 64)	Specifies the number of observations to send in a client/server transfer.
	TYPE= Data Set Option (p. 64)	Specifies the data set type for a specially structured SAS data set.
	WRITE= Data Set Option (p. 68)	Assigns a WRITE password to a SAS file that prevents users from writing to a file, unless they enter the password.
Miscellaneous	FILECLOSE= Data Set Option (p. 19)	Specifies how a tape is positioned when a SAS data set is closed.
Observation Control	FIRSTOBS= Data Set Option (p. 20)	Specifies the first observation that SAS processes in a SAS data set.
	IN= Data Set Option (p. 29)	Creates a Boolean variable that indicates whether the data set contributed data to the current observation.
	OBS= Data Set Option (p. 34)	Specifies the last observation that SAS processes in a data set.

Category	Language elements	Description
	POINTOBS= Data Set Option (p. 46)	Specifies whether SAS creates compressed data sets whose observations can be randomly accessed or sequentially accessed.
	WHERE= Data Set Option (p. 65)	Specifies specific conditions to use to select observations from a SAS data set.
	WHEREUP= Data Set Option (p. 67)	Specifies whether to evaluate new observations and modified observations against a WHERE expression.
User Control of SAS Index Usage	IDXNAME= Data Set Option (p. 25)	Directs SAS to use a specific index to match the conditions of a WHERE expression.
	IDXWHERE= Data Set Option (p. 26)	Specifies whether SAS uses an index search or a sequential search to match the conditions of a WHERE expression.
Variable Control	DROP= Data Set Option (p. 16)	For an input data set, excludes the specified variables from processing; for an output data set, excludes the specified variables from being written to the data set.
	KEEP= Data Set Option (p. 30)	For an input data set, specifies the variables to process; for an output data set, specifies the variables to write to the data set.
	RENAME= Data Set Option (p. 49)	Changes the name of a variable.

Dictionary

ALTER= Data Set Option

Assigns an ALTER password to a SAS file that prevents users from replacing or deleting the file, and enables access to a read- and write-protected file.

Valid in: DATA step and PROC steps

Category: Data Set Control

See: ALTER= Data Set Option under OpenVMS, UNIX, or z/OS in the documentation for your operating environment.

Syntax

ALTER=_alter-password_

Syntax Description

alter-password
   must be a valid SAS name. See Rules for Words and Names in the SAS Language .

Details

The ALTER= option applies to all types of SAS files except catalogs. You can use this option to assign a password to a SAS file or to access a read-protected, write-protected, or alter-protected SAS file.

When replacing a SAS data set that is protected with an ALTER password, the new data set inherits the ALTER password. To change the ALTER password for the new data set, use the MODIFY statement in the DATASETS procedure.

Note: A SAS password does not control access to a SAS file beyond the SAS system. You should use the operating system-supplied utilities and file-system security controls in order to control access to SAS files outside of SAS.

See Also

Data Set Options:

- "ENCRYPT= Data Set Option" on page 17
- "PW= Data Set Option" on page 47
- "READ= Data Set Option" on page 48
- "WRITE= Data Set Option" on page 68

Other:

- "File Protection" in Chapter 34 of *SAS Language Reference: Concepts*
- "Manipulating Passwords" in Chapter 15 of *Base SAS Procedures Guide*

BUFNO= Data Set Option

Specifies the number of buffers to be allocated for processing a SAS data set.

Valid in:	DATA step and PROC steps
Category:	Data Set Control
See:	BUFNO= Data Set Option in the documentation for your operating environment.

Syntax

BUFNO= *n* | *n*K | *hex*X | MIN | MAX

Syntax Description

n | *n*K

specifies the number of buffers in multiples of 1 (bytes); 1,024 (kilobytes). For example, a value of **8** specifies 8 buffers, and a value of **1k** specifies 1024 buffers.

*hex*X

specifies the number of buffers as a hexadecimal value. You must specify the value beginning with a number (0-9), followed by an X. For example, the value **2dx** sets the number of buffers to 45 buffers.

MIN

>   sets the minimum number of buffers to 0, which causes SAS to use the minimum optimal value for the operating environment. This is the default.

MAX

>   sets the number of buffers to the maximum possible number in your operating environment, up to the largest four-byte, signed integer, which is 2^{31}-1, or approximately 2 billion.

Details

The buffer number is not a permanent attribute of the data set; it is valid only for the current SAS session or job.

BUFNO= applies to SAS data sets that are opened for input, output, or update.

A larger number of buffers can speed up execution time by limiting the number of input and output (I/O) operations that are required for a particular SAS data set. However, the improvement in execution time comes at the expense of increased memory consumption.

To reduce I/O operations on a small data set as well as speed execution time, allocate one buffer for each page of data to be processed. This technique is most effective if you read the same observations several times during processing.

Operating Environment Information

>   The default value for BUFNO= is determined by your operating environment and is set to optimize sequential access. To improve performance for direct (random) access, you should change the value for BUFNO=. For the default setting and possible settings for direct access, see the BUFNO= data set option in the SAS documentation for your operating environment.

Comparisons

*   If the BUFNO= data set option is not specified, then the value of the BUFNO= system option is used. If both are specified in the same SAS session, the value specified for the BUFNO= data set option overrides the value specified for the BUFNO= system option.

*   To request that SAS allocate the number of buffers based on the number of data set pages and index file pages, use the SASFILE global statement.

See Also

Data Set Options:

*   "BUFSIZE= Data Set Option" on page 11

System Options:

*   "BUFNO= System Option" in *SAS System Options: Reference*

Statements:

*   "SASFILE Statement" in *SAS Statements: Reference*

BUFSIZE= Data Set Option

Specifies the size of a permanent buffer page for an output SAS data set.

Valid in:	DATA step and PROC steps
Category:	Data Set Control
Restriction:	Use with output data sets only.
See:	BUFSIZE= Data Set Option under UNIX, z/OS, or OpenVMS in the documentation for your operating environment.

Syntax

BUFSIZE= *n* | *n*K | *n*M | *n*G | *hex*X | MAX

Syntax Description

***n* | *n*K | *n*M | *n*G**

specifies the page size in multiples of 1 (bytes); 1,024 (kilobytes); 1,048,576 (megabytes); or 1,073,741,824 (gigabytes). For example, a value of **8** specifies a page size of 8 bytes, and a value of **4k** specifies a page size of 4096 bytes.

Note: If the system option and the data set option are not set, the default is 0. As a result, SAS uses the minimum optimal page size for the operating environment. The BUFSIZE= system option is used in either of the following scenarios:

- if the BUFSIZE= data set option is not set
- if the BUFSIZE= data set option is set to zero

Use BUFSIZE=0 in order to reset the buffer page size to the default value in your operating environment.

***hex*X**

specifies the page size as a hexadecimal value. You must specify the value beginning with a number (0-9), followed by an X. For example, the value **2dx** sets the page size to 45 bytes.

MAX

sets the page size to the maximum possible number in your operating environment, up to the largest four-byte, signed integer, which is $2^{31}-1$, or approximately 2 billion bytes.

Details

The page size is the amount of data that can be transferred for a single I/O operation to one buffer. The page size is a permanent attribute of the data set and is used when the data set is processed.

A larger page size can speed up execution time by reducing the number of times SAS has to read from or write to the storage medium. However, the improvement in execution time comes at the cost of increased memory consumption.

To change the page size, use a DATA step to copy the data set and either specify a new page or use the SAS default. To reset the page size to the default value in your operating environment, use BUFSIZE=0.

Note: If you use the COPY procedure to copy a data set to another library that is allocated with a different engine, the specified page size of the data set is not retained.

Operating Environment Information
The default value for BUFSIZE= is determined by your operating environment and is set to optimize sequential access. To improve performance for direct (random) access, you should change the value for BUFSIZE=. For the default setting and possible settings for direct access, see the BUFSIZE= data set option in the SAS documentation for your operating environment.

See Also

Data Set Options:

- "BUFNO= Data Set Option" on page 9

System Options:

- "BUFSIZE= System Option" in *SAS System Options: Reference*

CNTLLEV= Data Set Option

Specifies the level of shared access to a SAS data set.

Valid in:	DATA step and PROC steps
Category:	Data Set Control
Restriction:	Specify for input data sets only.

Syntax

CNTLLEV=LIB | MEM | REC

Syntax Description

LIB
specifies that concurrent access is controlled at the library level. Library-level control restricts concurrent access to only one update process to the library.

MEM
specifies that concurrent access is controlled at the SAS data set (member) level. Member-level control restricts concurrent access to only one update or output process to the SAS data set. If the data set is open for an update or output process, then no other operation can access the data set. If the data set is open for an input process, then other concurrent input processes are allowed but no update or output process is allowed.

REC
specifies that concurrent access is controlled at the observation (record) level. Record-level control allows more than one update access to the same SAS data set, but it denies concurrent update of the same observation.

Details

The CNTLLEV= option specifies the level at which shared update access to a SAS data set is denied. A SAS data set can be opened concurrently by more than one SAS session or by more than one statement, window, or procedure within a single session. By default, SAS procedures permit the greatest degree of concurrent access possible while they guarantee the integrity of the data and the data analysis. Therefore, you do not typically use the CNTLLEV= data set option.

Use this option when

- your application controls the access to the data, such as in SAS Component Language (SCL), SAS/IML software, or DATA step programming

- you access data through an interface engine that does not provide member-level control of the data.

If you use CNTLLEV=REC and the SAS procedure needs member-level control for integrity of the data analysis, SAS prints a warning to the SAS log. The warning states that inaccurate or unpredictable results can occur if the data are updated by another process during the analysis.

Example: Changing the Shared Access Level

In the following example, the first SET statement includes the CNTLLEV= data set option in order to override the default level of shared access from member-level control to record-level control. The second SET statement opens the SAS data set with the default member-level control.

```
set datalib.fuel (cntllev=rec) point=obsnum;
    .
    .
    .
set datalib.fuel;
    by area;
```

COMPRESS= Data Set Option

Specifies how observations are compressed in a new output SAS data set.

Valid in:	DATA step and PROC steps
Category:	Data Set Control
Restriction:	Use with output data sets only.

Syntax

COMPRESS=NO | YES | CHAR | BINARY

Syntax Description

NO
　　specifies that the observations in a newly created SAS data set are uncompressed (fixed-length records).

YES | CHAR

specifies that the observations in a newly created SAS data set are compressed (variable-length records) by SAS using RLE (Run Length Encoding). RLE compresses observations by reducing repeated consecutive characters (including blanks) to two-byte or three-byte representations.

Alias: ON

BINARY

specifies that the observations in a newly created SAS data set are compressed (variable-length records) by SAS using RDC (Ross Data Compression). RDC combines run-length encoding and sliding-window compression to compress the file.

Note: This method is highly effective for compressing medium to large (several hundred bytes or larger) blocks of binary data (numeric variables). Because the compression function operates on a single record at a time, the record length needs to be several hundred bytes or larger for effective compression.

Details

Compressing a file is a process that reduces the number of bytes required to represent each observation. Advantages of compressing a file include reduced storage requirements for the file and fewer I/O operations necessary to read or write to the data during processing. However, more CPU resources are required to read a compressed file (because of the overhead of uncompressing each observation). There are situations where the resulting file size might increase rather than decrease.

Use the COMPRESS= data set option to compress an individual file. Specify the option for output data sets only—that is, data sets named in the DATA statement of a DATA step or in the OUT= option of a SAS procedure. Use the COMPRESS= data set option only when you are creating a SAS data file (member type DATA). You cannot compress SAS views, because they contain no data.

After a file is compressed, the setting is a permanent attribute of the file, which means that to change the setting, you must re-create the file. That is, to uncompress a file, specify COMPRESS=NO for a DATA step that copies the compressed file.

Comparisons

The COMPRESS= data set option overrides the COMPRESS= option in the LIBNAME statement and the COMPRESS= system option.

The data set option POINTOBS=YES, which is the default, determines that a compressed data set can be processed with random access (by observation number) rather than sequential access. With random access, you can specify an observation number in the FSEDIT procedure and the POINT= option in the SET and MODIFY statements.

When you create a compressed file, you can also specify REUSE=YES (as a data set option or system option) in order to track and reuse space. With REUSE=YES, new observations are inserted in space freed when other observations are updated or deleted. When the default REUSE=NO is in effect, new observations are appended to the existing file.

POINTOBS=YES and REUSE=YES are mutually exclusive—that is, they cannot be used together. REUSE=YES takes precedence over POINTOBS=YES. That is, if you set REUSE=YES, SAS automatically sets POINTOBS=NO.

The TAPE engine supports the COMPRESS= data set option, but the engine does not support the COMPRESS= system option.

The XPORT engine does not support compression.

See Also

Data Set Options:

*   "POINTOBS= Data Set Option" on page 46
*   "REUSE= Data Set Option" on page 53

Statements:

*   "LIBNAME Statement" in *SAS Statements: Reference*

System Options:

*   "COMPRESS= System Option" in *SAS System Options: Reference*
*   "REUSE= System Option" in *SAS System Options: Reference*

Other:

*   "Compressing Data Files" in Chapter 26 of *SAS Language Reference: Concepts*

DLDMGACTION= Data Set Option

Specifies the action to take when a SAS data set in a SAS library is detected as damaged.

Valid in:	DATA step and PROC steps
Category:	Data Set Control

Syntax

DLDMGACTION=FAIL | ABORT | REPAIR | NOINDEX | PROMPT

Syntax Description

FAIL
> stops the step, issues an error message to the log immediately. This is the default for batch mode.

ABORT
> terminates the step, issues an error message to the log, and terminates the SAS session.

REPAIR
> automatically repairs and rebuilds indexes and integrity constraints, unless the data file is truncated. You use the REPAIR statement in PROC DATASETS to restore a truncated data set. It issues a warning message to the log. This is the default for interactive mode.

NOINDEX
> automatically repairs the data file without the indexes and integrity constraints, deletes the index file, updates the data file to reflect the disabled indexes and integrity constraints, and limits the data file to be opened only in INPUT mode. A warning is written to the SAS log instructing you to execute the PROC DATASETS REBUILD statement to correct or delete the disabled indexes and integrity constraints.

See:

"DLDMGACTION= Data Set Option" on page 15

"Recovering Disabled Indexes and Integrity Constraints" in Chapter 36 of *SAS Language Reference: Concepts*

PROMPT

displays a dialog box that asks you to select the FAIL, ABORT, REPAIR, or NOINDEX action.

DROP= Data Set Option

For an input data set, excludes the specified variables from processing; for an output data set, excludes the specified variables from being written to the data set.

Valid in:	DATA step and PROC steps
Category:	Variable Control

Syntax

DROP=*variable-1* <*...variable-n*>

Syntax Description

variable-1 <*..variable-n*>

lists one or more variable names. You can list the variables in any form that SAS allows.

Details

If the option is associated with an input data set, the variables are not available for processing. If the DROP= data set option is associated with an output data set, SAS does not write the variables to the output data set, but they are available for processing.

Comparisons

- The DROP= data set option differs from the DROP statement in these ways:

  - In DATA steps, the DROP= data set option can apply to both input and output data sets. The DROP statement applies only to output data sets.

  - In DATA steps, when you create multiple output data sets, use the DROP= data set option to write different variables to different data sets. The DROP statement applies to all output data sets.

  - In PROC steps, you can use only the DROP= data set option, not the DROP statement.

- The KEEP= data set option specifies a list of variables to be included in processing or to be written to the output data set.

Examples

Example 1: Excluding Variables from Input

In this example, the variables SALARY and GENDER are not included in processing and they are not written to either output data set:

```
data plan1 plan2;
   set payroll(drop=salary gender);
   if hired<'01jan98'd then output plan1;
   else output plan2;
run;
```

You cannot use SALARY or GENDER in any logic in the DATA step because DROP= prevents the SET statement from reading them from PAYROLL.

Example 2: Processing Variables without Writing Them to a Data Set

In this example, SALARY and GENDER are not written to PLAN2, but they are written to PLAN1:

```
data plan1 plan2(drop=salary gender);
   set payroll;
   if hired<'01jan98'd then output plan1;
   else output plan2;
run;
```

See Also

Data Set Options:

- "KEEP= Data Set Option" on page 30

Statements:

- "DROP Statement" in *SAS Statements: Reference*

ENCRYPT= Data Set Option

Specifies whether to encrypt an output SAS data set.

Valid in:	DATA step and PROC steps
Category:	Data Set Control
Restriction:	Use with output data sets only.

Syntax

ENCRYPT=YES | NO

Syntax Description

YES

> encrypts the file. This encryption uses passwords that are stored in the data set. At a minimum, you must specify the READ= or the PW= data set option at the same time

that you specify ENCRYPT=YES. Because the encryption method uses passwords, you cannot change *any* password on an encrypted data set without recreating the data set.

CAUTION:

> **Record all passwords when using ENCRYPT=YES.** If you forget the passwords, you cannot reset it without assistance from SAS. The process is time-consuming and resource-intensive.

NO

does not encrypt the file.

Details

When using ENCRYPT=YES, the following rules apply:

- You cannot encrypt SAS views, because they contain no data.

- To copy an encrypted data file, the output engine must support the encryption. Otherwise, the data file is not copied.

- Encrypted files work only in SAS 6.11 or later.

- If the data file is encrypted, all associated indexes are also encrypted.

- Encryption requires approximately the same amount of CPU resources as compression.

- You cannot use PROC CPORT on SAS Proprietary encrypted data files.

Example: Using the ENCRYPT=YES Option

This example creates an encrypted SAS data set using encryption:

```
data salary(encrypt=yes read=green);
   input name $ yrsal bonuspct;
   datalines;
Muriel    34567   3.2
Bjorn     74644   2.5
Freda     38755   4.1
Benny     29855   3.5
Agnetha   70998   4.1
;
```

To use this data set, specify the read password:

```
proc contents data=salary(read=green);
run;
```

See Also

Data Set Options:

- "ALTER= Data Set Option" on page 8

- "PW= Data Set Option" on page 47

- "READ= Data Set Option" on page 48

- "WRITE= Data Set Option" on page 68

Other:

- "SAS Data File Encryption" in Chapter 34 of *SAS Language Reference: Concepts*

EXTENDOBSCOUNTER= Data Set Option

Specifies whether to extend the maximum observation count in a new output SAS data file.

Valid in:	DATA step and PROC steps
Category:	Data Set Control
Alias:	EOC=
Default:	NO
Restrictions:	Use with output data files only.
	Use with the BASE engine only.

Syntax

EXTENDOBSCOUNTER=NO | YES

Syntax Description

NO

specifies that the maximum observation count in a newly created SAS data file is determined by the long integer size for the operating environment. In operating environments with a 32-bit long integer, the maximum number is $2^{31}-1$ or approximately two billion observations (2,147,483,647). In operating environments with a 64-bit long integer, the number is $2^{63}-1$ or approximately 9.2 quintillion observations.

YES

requests an enhanced file format in a newly created SAS data file that counts observations beyond the 32-bit long limitation. For a SAS data file that is created for an operating environment that stores the number of observations with a 32-bit long integer, the data file behaves like a 64-bit file with respect to counters.

Restrictions:

A SAS data file that is created with EXTENDOBSCOUNTER=YES is incompatible with releases before SAS 9.3.

Specify EXTENDOBSCOUNTER=YES only for an output SAS data file whose internal data representation stores the observation count as a 32-bit long integer. For a table that lists the operating environments and OUTREP= data representation values that are appropriate with EXTENDOBSCOUNTER=YES, see "When to Use the EXTENDOBSCOUNTER=YES Option" in Chapter 26 of *SAS Language Reference: Concepts*.

See Also

"Extending the Observation Count in a SAS Data File" in Chapter 26 of *SAS Language Reference: Concepts*

FILECLOSE= Data Set Option

Specifies how a tape is positioned when a SAS data set is closed.

Valid in:	DATA step and PROC steps
Category:	Miscellaneous
Restriction:	
See:	FILECLOSE= Data Set Option under in the documentation UNIX operating environment.
CAUTION:	**The option values are not recognized by all operating environments.** Additional values are available on some operating environments. See the appropriate sections of the SAS documentation for your operating environment for more information about using SAS libraries that are stored on tape.

Syntax

FILECLOSE=DISP | LEAVE | REREAD | REWIND

Syntax Description

DISP

positions the tape volume according to the disposition specified in the operating environment's control language.

LEAVE

positions the tape at the end of the file that was just processed. Use FILECLOSE=LEAVE if you are not repeatedly accessing the same files in a SAS program but you are accessing one or more subsequent SAS files on the same tape.

REREAD

positions the tape volume at the beginning of the file that was just processed. Use FILECLOSE=REREAD if you are accessing the same SAS data set on tape several times in a SAS program.

REWIND

rewinds the tape volume to the beginning. Use FILECLOSE=REWIND if you are accessing one or more previous SAS files on the same tape, but you are not repeatedly accessing the same files in a SAS program.

FIRSTOBS= Data Set Option

Specifies the first observation that SAS processes in a SAS data set.

Valid in:	DATA step and PROC steps
Category:	Observation Control
Restrictions:	Valid for input (read) processing only.
	Cannot use with PROC SQL views.

Syntax

FIRSTOBS= *n*| *n*K | *n*M | *n*G | *hex*X | MIN | MAX

Syntax Description

n | *n*K | *n*M | *n*G

specifics the number of the first observation to process in multiples of 1 (bytes); 1,024 (kilobytes); 1,048,576 (megabytes); or 1,073,741,824 (gigabytes). For example, a value of **8** specifies the 8th observation, and a value of **3k** specifies 3,072.

*hex*X

specifies the number of the first observation to process as a hexadecimal value. You must specify the value beginning with a number (0-9), followed by an X. For example, the value **2dx** sets the 45th observation as the first observation to process.

MIN

sets the number of the first observation to process to 1. This is the default.

MAX

sets the number of the first observation to process to the maximum number of observations in the data set, up to the largest eight-byte, signed integer, which is $2^{63}-1$, or approximately 9.2 quintillion observations.

Details

The FIRSTOBS= data set option affects a single, existing SAS data set. Use the FIRSTOBS= system option to affect all steps for the duration of your current SAS session.

FIRSTOBS= is valid for input (read) processing only. Specifying FIRSTOBS= is not valid for output or update processing.

You can apply FIRSTOBS= processing to WHERE processing. For more information, see "Processing a Segment of Data That Is Conditionally Selected" in Chapter 11 of *SAS Language Reference: Concepts* .

Comparisons

* The FIRSTOBS= data set option overrides the FIRSTOBS= system option for the individual data set.

* While the FIRSTOBS= data set option specifies a starting point for processing, the OBS= data set option specifies an ending point. The two options are often used together to define a range of observations to be processed.

* When external files are read, the FIRSTOBS= option in the INFILE statement specifies which record to read first.

Example: Example

This PROC step prints the data set STUDY beginning with observation 20:

```
proc print data=study(firstobs=20);
run;
```

This SET statement uses both FIRSTOBS= and OBS= to read-only observations 5 through 10 from the data set STUDY. Data set NEW contains six observations.

```
data new;
  set study(firstobs=5 obs=10);
run;
```

```
proc print data=new;
 run;
```

See Also

Data Set Options:

- "OBS= Data Set Option" on page 34

Statements:

- "INFILE Statement" in *SAS Statements: Reference*
- "WHERE Statement" in *SAS Statements: Reference*

System Options:

- "FIRSTOBS= System Option" in *SAS System Options: Reference*

GENMAX= Data Set Option

Requests generations for a new data set, modifies the number of generations for an existing data set, and specifies the maximum number of versions.

Valid in:	DATA step and PROC steps
Category:	Data Set Control
Restriction:	Use with output data sets only.

Syntax

GENMAX=*number-of-generations*

Syntax Description

number-of-generations
    requests generations for a data set and specifies the maximum number of versions to maintain. The value can be from 0 to 1000. The default is GENMAX=0, which means that no generation data sets are requested.

Details

You use GENMAX= to request generations for a new data set and to modify the number of generations for an existing data set. The first time the data set is replaced, SAS keeps the replaced version and appends a four-character version number to its member name, which includes # and a three-digit number. For example, for a data set named A, a historical version would be A#001.

After generations of a data set are requested, the member name is limited to 28 characters (rather than 32). The last four characters are reserved for the appended version number. When the GENMAX= data set option is set to 0, the member name can be up to 32 characters.

If you reduce the number of generations for an existing data set, SAS deletes the oldest versions above the new limit.

Examples

Example 1: Requesting Generations When You Create a Data Set

This example shows how to request generations for a new data set. The DATA step creates a data set named WORK.A that can have as many as 10 generations (one current version and nine historical versions):

```
data a(genmax=10);
   x=1;
   output;
run;
```

Example 2: Modifying the Number of Generations on an Existing Data Set

This example shows how to change the number of generations on the data set MYLIB.A to 4:

```
proc datasets lib=mylib;
   modify a(genmax=4);
run;
```

See Also

Data Set Option:

- "GENNUM= Data Set Option" on page 23

Other:

- "Understanding Generation Data Sets" in Chapter 26 of *SAS Language Reference: Concepts*

GENNUM= Data Set Option

Specifies a particular generation of a SAS data set.

Valid in:	DATA step and PROC steps
Category:	Data Set Control
Restriction:	Use with input data sets only.

Syntax

GENNUM=*integer*

Syntax Description

integer

is a number that references a specific version from a generation group. Specifying a positive number is an absolute reference to a specific generation number that is appended to a data set's name. Specifying a negative number is a relative reference to a historical version in relation to the base version, from the youngest to the oldest. Typically, a value of 0 refers to the current (base) version.

The DATASETS procedure provides a variety of statements for which specifying GENNUM= has additional functionality:

- For the DATASETS and DELETE statements, GENNUM= supports the additional values ALL, HIST, and REVERT.

- For the CHANGE statement, GENNUM= supports the additional value ALL.

- For the CHANGE statement, specifying GENNUM=0 refers to all versions rather than just the base version.

Details

After generations for a data set have been requested using the GENMAX= data set option, use GENNUM= to request a specific version. For example, specifying GENNUM=3 refers to the historical version #003, while specifying GENNUM=-1 refers to the youngest historical version.

Note that after 999 replacements, the youngest version would be #999. After 1,000 replacements, SAS rolls over the youngest version number to #000. Therefore, if you want the historical version #000, specify GENNUM=1000.

Both an absolute reference and a relative reference refer to a specific version. A relative reference does not skip deleted versions. Therefore, when working with a generation group that includes one or more deleted versions, using a relative reference results in an error if the version being referenced has been deleted. For example, if you have the base version AIR and three historical versions (AIR#001, AIR#002, and AIR#003) and you delete AIR#002, the following statements return an error, because AIR#002 does not exist. SAS does not assume you mean AIR#003:

```
proc print data=air (gennum= -2);
run;
```

Examples

Example 1: Requesting a Version Using an Absolute Reference

This example prints the historical version #003 for data set A, using an absolute reference:

```
proc print data=a(gennum=3);
run;
```

Example 2: Requesting A Version Using a Relative Reference

The following PRINT procedure prints the data set three versions back from the base version:

```
proc print data=a(gennum=-3);
run;
```

See Also

Data Set Option:

- "GENMAX= Data Set Option" on page 22

Other:

- "Understanding Generation Data Sets" in Chapter 26 of *SAS Language Reference: Concepts*
- Chapter 15, "DATASETS Procedure" in *Base SAS Procedures Guide*

IDXNAME= Data Set Option

Directs SAS to use a specific index to match the conditions of a WHERE expression.

Valid in:	DATA step and PROC steps
Category:	User Control of SAS Index Usage
Restrictions:	Use with input data sets only
	Mutually exclusive with IDXWHERE= data set option

Syntax

IDXNAME=*index-name*

Syntax Description

index-name

specifies the name (up to 32 characters) of a simple or composite index for the SAS data set. SAS does not attempt to determine whether the specified index is the best one or whether a sequential search might be more resource efficient.

Interaction: The specification is not a permanent attribute of the data set and is valid only for the current use of the data set.

Tip: To request that IDXNAME= usage be noted in the SAS log, specify the system option MSGLEVEL=I.

Details

By default, to satisfy the conditions of a WHERE expression for an indexed SAS data set, SAS identifies zero or more candidate indexes that could be used to optimize the WHERE expression. From the list of candidate indexes, SAS determines the one that provides the best performance, or rejects all of the indexes if a sequential pass of the data is expected to be more efficient.

Because the index SAS selects might not always provide the best optimization, you can direct SAS to use one of the candidate indexes by specifying the IDXNAME= data set option. If you specify an index that SAS does not identify as a candidate index, then IDXNAME= does not process the request. That is, IDXNAME= does not allow you to specify an index that would produce incorrect results.

Comparisons

IDXWHERE= enables you to override the SAS decision about whether to use an index.

Example: Example

This example uses the IDXNAME= data set option in order to direct SAS to use a specific index to optimize the WHERE expression. SAS then disregards the possibility that a sequential search of the data set might be more resource efficient and does not attempt to determine whether the specified index is the best one. (Note that the EMPNUM index was not created with the NOMISS option.)

```
data mydata.empnew;
   set mydata.employee (idxname=empnum);
   where empnum < 2000;
run;
```

See Also

Data Set Option:

- "IDXWHERE= Data Set Option" on page 26

Other:

- "Using an Index for WHERE Processing" in Chapter 26 of *SAS Language Reference: Concepts*

IDXWHERE= Data Set Option

Specifies whether SAS uses an index search or a sequential search to match the conditions of a WHERE expression.

Valid in:	DATA step and PROC steps
Category:	User Control of SAS Index Usage
Restrictions:	Use with input data sets only.
	Mutually exclusive with IDXNAME= data set option

Syntax

IDXWHERE=YES | NO

Syntax Description

YES

> tells SAS to choose the best index to optimize a WHERE expression, and to disregard the possibility that a sequential search of the data set might be more resource-efficient.

NO

> tells SAS to ignore all indexes and satisfy the conditions of a WHERE expression with a sequential search of the data set.

> *Note:* You cannot use IDXWHERE= to override the use of an index to process a BY statement.

Details

By default, to satisfy the conditions of a WHERE expression for an indexed SAS data set, SAS decides whether to use an index or to read the data set sequentially. The software estimates the relative efficiency and chooses the method that is more efficient.

You might need to override the software's decision by specifying the IDXWHERE= data set option because the decision is based on general rules that might occasionally not produce the best results. That is, by specifying the IDXWHERE= data set option, you are able to determine the processing method.

Note: The specification is not a permanent attribute of the data set and is valid only for the current use of the data set.

Note: If you issue the system option MSGLEVEL=I, you can request that IDXWHERE= usage be noted in the SAS log if the setting affects index processing.

Comparisons

IDXNAME= enables you to direct SAS to use a specific index.

Examples

Example 1: Specifying Index Usage

This example uses the IDXWHERE= data set option to tell SAS to decide which index is the best to optimize the WHERE expression. SAS then disregards the possibility that a sequential search of the data set might be more resource-efficient:

```
data mydata.empnew;
    set mydata.employee (idxwhere=yes);
    where empnum < 2000;
```

Example 2: Specifying No Index Usage

This example uses the IDXWHERE= data set option to tell SAS to ignore any index and to satisfy the conditions of the WHERE expression with a sequential search of the data set:

```
data mydata.empnew;
    set mydata.employee (idxwhere=no);
    where empnum < 2000;
```

See Also

Data Set Option:

- "IDXNAME= Data Set Option" on page 25

Other:

- "Understanding SAS Indexes" in Chapter 26 of *SAS Language Reference: Concepts*
- WHERE-Expression Processing

INDEX= Data Set Option

Defines an index for a new output SAS data set.

Valid in:	DATA step and PROC steps
Category:	Data Set Control
Restriction:	Use with output data sets only.

Syntax

INDEX=(*index-specification-1* ...<*index-specification-n*>)

Syntax Description

index-specification
> names and describes a simple or a composite index to be built. *Index-specification* has this form:
>
> *index= (variable(s)* </UNIQUE> </NOMISS>)

index
> is the name of a variable that forms the index or the name that you choose for a composite index.

variable or variables
> is a list of variables to use in making a composite index.

UNIQUE
> specifies that the values of the key variables must be unique. If you specify UNIQUE for a new data set and multiple observations have the same values for the index variables, the index is not created. A slash (/) must precede the UNIQUE option.

NOMISS
> excludes all observations with missing values from the index. Observations with missing values are still read from the data set but not through the index. A slash (/) must precede the NOMISS option.

Examples

Example 1: Defining a Simple Index
The following INDEX= data set option defines a simple index for the SSN variable:

```
data new(index=(ssn));
```

Example 2: Defining a Composite Index
The following INDEX= data set option defines a composite index named CITYST that uses the CITY and STATE variables:

```
data new(index=(cityst=(city state)));
```

Example 3: Defining a Simple and a Composite Index
The following INDEX= data set option defines a simple index for SSN and a composite index for CITY and STATE:

```
data new(index=(ssn cityst=(city state)));
```

Example 4: Defining a Simple Index with the UNIQUE Option
The following INDEX= data set option defines a simple index for the SSN variable with unique values:

```
data new(index=(ssn /unique));
```

Example 5: Defining a Simple Index with the NOMISS Option
The following INDEX= data set option defines a simple index for the SSN variable, excluding all observations with missing values from the index:

```
data new(index=(ssn /nomiss));
```

Example 6: Defining Multiple Indexes Using the UNIQUE and NOMISS Options

The following INDEX= data set option defines a simple index for the SSN variable and a composite index for CITY and STATE. Each variable must have a UNIQUE and NOMISS option:

```
data new(index=(ssn /unique/nomiss cityst=(city state)/unique/nomiss));
```

See Also

- XisError: User must supply overrideFetchedText because link target element has no title

- XisError: User must supply overrideFetchedText because link target element has no title

- "Understanding SAS Indexes" in Chapter 26 of *SAS Language Reference: Concepts*

IN= Data Set Option

Creates a Boolean variable that indicates whether the data set contributed data to the current observation.

Valid in:	DATA step
Category:	Observation Control
Restriction:	Use with the SET, MERGE, MODIFY, and UPDATE statements only.

Syntax

IN=*variable*

Syntax Description

variable
> names the new variable whose value indicates whether that input data set contributed data to the current observation. Within the DATA step, the value of the variable is 1 if the data set contributed to the current observation, and 0 otherwise.

Details

Specify the IN= data set option in parentheses after a SAS data set name in the SET, MERGE, MODIFY, and UPDATE statements only. Values of IN= variables are available to program statements during the DATA step. The variables are not included in the SAS data set that is being created, unless they are assigned to a new variable.

When you use IN= with BY–group processing, and when a data set contributes an observation for the current BY group, the IN= value is 1. The value remains as long as that BY group is still being processed and the value is not reset by programming logic.

Example

In this example, IN= creates a new variable, OVERSEAS, that denotes international flights. The variable I has a value of 1 when the observation is read from the NONUSA data set. Otherwise, it has a value of 0. The IF-THEN statement checks the value of I to determine whether the data set NONUSA contributed data to the current observation. If I=1, the variable OVERSEAS receives an asterisk (*) as a value.

```
data allflts;
    set usa nonusa(in=i);
    by fltnum;
    if i then overseas='*';
run;
```

See Also

Statements:

- "BY Statement" in *SAS Statements: Reference*

- "MERGE Statement" in *SAS Statements: Reference*

- "MODIFY Statement" in *SAS Statements: Reference*

- "SET Statement" in *SAS Statements: Reference*

- "UPDATE Statement" in *SAS Statements: Reference*

Other:

- BY-GROUP Processing

KEEP= Data Set Option

For an input data set, specifies the variables to process; for an output data set, specifies the variables to write to the data set.

Valid in:	DATA step and PROC steps
Category:	Variable Control

Syntax

KEEP=*variable-1* <*...variable-n*>

Syntax Description

variable-1 <*...variable-n*>
   lists one or more variable names. You can list the variables in any form that SAS allows.

Details

If the KEEP= data set option is associated with an input data set, only those variables that are listed after the KEEP= data set option are available for processing. If the KEEP= data set option is associated with an output data set, only the variables listed after the option are written to the output data set. All variables are available for processing.

Comparisons

- The KEEP= data set option differs from the KEEP statement in the following ways:

- In DATA steps, the KEEP= data set option can apply to both input and output data sets. The KEEP statement applies only to output data sets.

- In DATA steps, when you create multiple output data sets, use the KEEP= data set option to write different variables to different data sets. The KEEP statement applies to all output data sets.

- In PROC steps, you can use only the KEEP= data set option, not the KEEP statement.

- The DROP= data set option specifies variables to omit during processing or to omit from the output data set.

Example

In this example, only IDNUM and SALARY are read from PAYROLL, and they are the only variables in PAYROLL that are available for processing:

```
data bonus;
   set payroll(keep=idnum salary);
   bonus=salary*1.1;
run;
```

See Also

Data Set Options:

- "DROP= Data Set Option" on page 16

Statements:

- "KEEP Statement" in *SAS Statements: Reference*

LABEL= Data Set Option

Specifies a label for a SAS data set.

Valid in:	DATA step and PROC steps
Category:	Data Set Control

Syntax

LABEL='*label*'

Syntax Description

'*label*'

specifies a text string of up to 256 characters. If the label text contains single quotation marks, use double quotation marks around the label. You also can use two single quotation marks in the label text and surround the string with single quotation marks. To remove a label from a data set, assign a label that is equal to a blank that is enclosed in quotation marks.

Details

You can use the LABEL= option on both input and output data sets. When you use LABEL= on input data sets, it assigns a label for the file for the duration of that DATA or PROC step. When it is specified for an output data set, the label becomes a permanent part of that file. The file can be printed using the CONTENTS or DATASETS procedure, and modified using PROC DATASETS.

A label assigned to a data set remains associated with that data set when you update a data set in place, such as using the APPEND procedure or the MODIFY statement. However, a label is lost if you use a data set with a previously assigned label to create a new data set in the DATA step. For example, a label previously assigned to data set ONE is lost when you create the new output data set ONE in this DATA step:

```
data one;
     set one;
run;
```

Comparisons

- The LABEL= data set option enables you to specify labels only for data sets. You can specify labels for the variables in a data set using the LABEL statement.

- The LABEL= option in the ATTRIB statement also enables you to assign labels to variables.

Example

These examples assign labels to data sets:

```
data w2(label='1976 W2 Info, Hourly');

data new(label='Peter''s List');

data new(label="Hillside's Daily Account");

data sales(label='Sales For May(NE)');
```

See Also

Statements:

- "ATTRIB Statement" in *SAS Statements: Reference*
- "LABEL Statement" in *SAS Statements: Reference*
- "MODIFY Statement" in *SAS Statements: Reference*

Procedures:

- Chapter 12, "CONTENTS Procedure" in *Base SAS Procedures Guide*
- Chapter 15, "DATASETS Procedure" in *Base SAS Procedures Guide*

OBSBUF= Data Set Option

Determines the size of the view buffer for processing a DATA step view.

Valid in: DATA step and PROC steps

Category:	Data Set Control
Restriction:	Valid only for a DATA step view

Syntax

OBSBUF=n

Syntax Description

n

specifies the number of observations that are read into the view buffer at a time.

Default: 32K bytes of memory are allocated for the default view buffer, which means that the default number of observations that can be read into the view buffer at one time depends on the observation length. Therefore, the default is the number of observations that can fit into 32K bytes. If the observation length is larger than 32K, then only one observation can be read into the buffer at a time.

Tip: To determine the observation length, which is its size in bytes, use PROC CONTENTS for the DATA step view.

CAUTION: The maximum value for the OBSBUF= option depends on the amount of available memory. If you specify a value so large that the memory allocation of the view buffer fails, an out-of- memory error results. If you specify a value that is larger than the amount of available real memory and your operating environment allows SAS to perform the allocation using virtual memory, the result can be a decrease in performance due to excessive paging.

Details

The OBSBUF= data set option specifies the number of observations that can be read into the view buffer at a time. The *view buffer* is a segment of memory that is allocated to hold output observations that are generated from a DATA step view. The size of the buffer determines how much data can be held in memory at one time. OBSBUF= enables you to tune the performance of reading data from a DATA step view.

The view buffer is shared between the request that opens the DATA step view (for example, a SAS procedure) and the DATA step view itself. Two computer tasks coordinate between requesting data and generating and returning the data as follows:

1. When a request task (such as a PRINT procedure) requests data, task switching occurs from the request task to the view task in order to execute the DATA step view and generate the observations. The DATA step view fills the view buffer with as many observations as possible.

2. When the view buffer is full, task switching occurs from the view task back to the request task in order to return the requested data. The observations are cleared from the view buffer.

The size of the view buffer determines how many generated observations can be held. The number of generated observations then determines how many times the computer must switch between the request task and the view task. For example, OBSBUF=1 results in task switching for each observation, while OBSBUF=10 results in 10 observations being read into the view buffer at a time. The larger the view buffer is, the less task switching is needed to process a DATA step view, which can speed up execution time.

To improve efficiency, first determine how many observations fits into the default buffer size, then set the view buffer so that it can hold more generated observations.

Note: Using OBSBUF= can improve processing efficiency by reducing task switching. However, the larger the view buffer size, the more time it takes to fill. This delays the task switching from the view task back to the request task in order to return the requested data. The delay is more apparent in interactive applications. For example, when you use the Viewtable window, the larger the view buffer, the longer it takes to display the requested observations. The view buffer must be filled before even one observation is returned to the Viewtable. Therefore, before you set a very large view buffer size, consider the type of application that you are using to process the DATA step view as well as the amount of memory that you have available.

Example

For this example, the observation length is 10K, which means that the default view buffer size, which is 32K, would result in three observations at a time to be read into the view buffer. The default view buffer size causes the execution time to be slower, because the computer must do task switching for every three observations that are generated.

To improve performance, the OBSBUF= data set option is set to 100. This action causes one hundred observations at a time to be read into the view buffer. It also reduces task switching in order to process the DATA step view with the PRINT procedure:

```
data testview / view=testview;
   ... more SAS statements ...
run;
proc print data=testview (obsbuf=100);
run;
```

See Also

Data Set Options:

- "SPILL= Data Set Option" on page 57

OBS= Data Set Option

Specifies the last observation that SAS processes in a data set.

Valid in:	DATA step and PROC steps
Category:	Observation Control
Default:	MAX
Restrictions:	Use with input data sets only
	Cannot use with PROC SQL views

Syntax

OBS= n | nK | nM | nG | nT | *hex*X | MIN | MAX

Syntax Description

n | *n*K | *n*M | *n*G | *n*T

specifies a number to indicate when to stop processing observations, with *n* being an integer. Using one of the letter notations results in multiplying the integer by a specific value. That is, specifying K (kilo) multiplies the integer by 1,024, M (mega) multiplies by 1,048,576, G (giga) multiplies by 1,073,741,824, or T (tera) multiplies by 1,099,511,627,776. For example, a value of **20** specifies 20 observations, while a value of **3m** specifies 3,145,728 observations.

*hex*X

specifies a number to indicate when to stop processing as a hexadecimal value. You must specify the value beginning with a number (0–9), followed by an X. For example, the hexadecimal value F8 must be specified as **0F8x** in order to specify the decimal equivalent of 248. The value **2dx** specifies the decimal equivalent of 45.

MIN

sets the number to indicate when to stop processing to 0. Use OBS=0 in order to create an empty data set that has the structure, but not the observations, of another data set.

Interaction: If OBS=0 and the NOREPLACE option is in effect, then SAS can still take certain actions. SAS actually executes each DATA and PROC step in the program, using no observations. For example, SAS executes procedures, such as CONTENTS and DATASETS, that process libraries or SAS data sets.

MAX

sets the number to indicate when to stop processing to the maximum number of observations in the data set, up to the largest 8-byte, signed integer, which is $2^{63}-1$, or approximately 9.2 quintillion. This is the default.

Details

OBS= tells SAS when to stop processing observations. To determine when to stop processing, SAS uses the value for OBS= in a formula that includes the value for OBS= and the value for FIRSTOBS=. The formula is

```
(obs - firstobs) + 1 = results
```

For example, if OBS=10 and FIRSTOBS=1 (which is the default for FIRSTOBS=), the result is ten observations, that is `(10 - 1) + 1 = 10`. If OBS=10 and FIRSTOBS=2, the result is nine observations, that is `(10 - 2) + 1 = 9`. OBS= is valid only when an existing SAS data set is read.

Comparisons

* The OBS= data set option overrides the OBS= system option for the individual data set.

* While the OBS= data set option specifies an ending point for processing, the FIRSTOBS= data set option specifies a starting point. The two options are often used together to define a range of observations to be processed.

* The OBS= data set option enables you to select observations from SAS data sets. You can select observations to be read from external data files by using the OBS= option in the INFILE statement.

Examples

Example 1: Using OBS= to Specify When to Stop Processing Observations

This example illustrates the result of using OBS= to tell SAS when to stop processing observations. This example creates a SAS data set and executes the PRINT procedure with FIRSTOBS=2 and OBS=12. The result is 11 observations, that is `(12 - 2) + 1 = 11`. The result of OBS= in this situation appears to be the observation number that SAS processes last, because the output starts with observation 2, and ends with observation 12. This situation is only a coincidence.

```
data Ages;
   input Name $ Age;
   datalines;
Miguel 53
Brad 27
Willie 69
Marc 50
Sylvia 40
Arun 25
Gary 40
Becky 51
Alma 39
Tom 62
Kris 66
Paul 60
Randy 43
Barbara 52
Virginia 72
;
proc print data=Ages (firstobs=2 obs=12);
run;
```

Display 2.1 *PROC PRINT Output Using OBS= and FIRSTOBS=*

The SAS System

Obs	Name	Age
2	Brad	27
3	Willie	69
4	Marc	50
5	Sylvia	40
6	Arun	25
7	Gary	40
8	Becky	51
9	Alma	39
10	Tom	62
11	Kris	66
12	Paul	60

Example 2: Using OBS= with WHERE Processing

This example illustrates the result of using OBS= along with WHERE processing. The example uses the data set that was created in Example 1, which contains 15 observations.

First, here is the PRINT procedure with a WHERE statement. The subset of the data results in 12 observations:

```
proc print data=Ages;
   where Age LT 65;
run;
```

Display 2.2 *PROC PRINT Output Using a WHERE Statement*

The SAS System

Obs	Name	Age
1	Miguel	53
2	Brad	27
4	Marc	50
5	Sylvia	40
6	Arun	25
7	Gary	40
8	Becky	51
9	Alma	39
10	Tom	62
12	Paul	60
13	Randy	43
14	Barbara	52

Executing the PRINT procedure with the WHERE statement and OBS=10 results in 10 observations, that is `(10 - 1) + 1 = 10`. Note that with WHERE processing, SAS first subsets the data and applies OBS= to the subset.

```
proc print data=Ages (obs=10);
   where Age LT 65;
run;
```

Display 2.3 *PROC PRINT Output Using a WHERE Statement and OBS=*

The SAS System

Obs	Name	Age
1	Miguel	53
2	Brad	27
4	Marc	50
5	Sylvia	40
6	Arun	25
7	Gary	40
8	Becky	51
9	Alma	39
10	Tom	62
12	Paul	60

The result of OBS= appears to be how many observations to process, because the output consists of 10 observations, ending with the observation number 12. However, the result is only a coincidence. If you apply FIRSTOBS=2 and OBS=10 to the subset, then the result is nine observations, that is **(10 - 2) + 1 = 9**. OBS= in this situation is neither the observation number to end with nor how many observations to process; the value is used in the formula to determine when to stop processing.

```
proc print data=Ages (firstobs=2 obs=10);
   where Age LT 65;
run;
```

Display 2.4 *PROC PRINT Output Using WHERE Statement, OBS=, and FIRSTOBS=*

Obs	Name	Age
2	Brad	27
4	Marc	50
5	Sylvia	40
6	Arun	25
7	Gary	40
8	Becky	51
9	Alma	39
10	Tom	62
12	Paul	60

The SAS System

This example illustrates the result of using OBS= for a data set that has deleted observations. The example uses the data set that was created in Example 1, with observation 6 deleted.

First, here is PROC PRINT output of the modified file:

```
proc print data=Ages;
run;
```

Display 2.5 *PROC PRINT Output Showing Observation 6 Deleted*

The SAS System

Obs	Name	Age
1	Miguel	53
2	Brad	27
3	Willie	69
4	Marc	50
5	Sylvia	40
7	Gary	40
8	Becky	51
9	Alma	39
10	Tom	62
11	Kris	66
12	Paul	60
13	Randy	43
14	Barbara	52
15	Virginia	72

Executing the PRINT procedure with OBS=12 results in 12 observations, that is `(12 - 1) + 1 = 12`:

```
proc print data=Ages (obs=12);
run;
```

Display 2.6 PROC PRINT Output Using OBS=

Obs	Name	Age
1	Miguel	53
2	Brad	27
3	Willie	69
4	Marc	50
5	Sylvia	40
7	Gary	40
8	Becky	51
9	Alma	39
10	Tom	62
11	Kris	66
12	Paul	60
13	Randy	43

The SAS System

Example 3: Using OBS= When Observations Are Deleted

The result of OBS= appears to be how many observations to process, because the output consists of 12 observations, ending with the observation number 13. However, if you apply FIRSTOBS=2 and OBS=12, the result is 11 observations, that is **(12 - 2) + 1 = 11**. OBS= in this situation is neither the observation number to end with nor how many observations to process; the value is used in the formula to determine when to stop processing.

```
proc print data=Ages (firstobs=2 obs=12);
run;
```

Display 2.7 PROC PRINT Output Using OBS= and FIRSTOBS=

The SAS System

Obs	Name	Age
2	Brad	27
3	Willie	69
4	Marc	50
5	Sylvia	40
7	Gary	40
8	Becky	51
9	Alma	39
10	Tom	62
11	Kris	66
12	Paul	60
13	Randy	43

See Also

Data Set Options:

- "FIRSTOBS= Data Set Option" on page 20

Statements:

- "INFILE Statement" in *SAS Statements: Reference*
- "WHERE Statement" in *SAS Statements: Reference*
- *SAS Statements: Reference*

System Options:

- "OBS= System Option" in *SAS System Options: Reference*

For more information about using OBS= and WHERE processing, see

- "Processing a Segment of Data That Is Conditionally Selected" in Chapter 11 of *SAS Language Reference: Concepts*

OUTREP= Data Set Option

Specifies the data representation for the output SAS data set.

Valid in:	DATA step and PROC steps
Category:	Data Set Control
See:	OUTREP= data set option in the *SAS Companion for z/OS*.

Syntax

OUTREP=	*format*

Syntax Description

format

specifies the data representation, which is the form in which data is stored in a particular operating environment. Different operating environments use different standards or conventions for storing floating-point numbers (for example, IEEE or IBM mainframe); for character encoding (ASCII or EBCDIC); for the ordering of bytes in memory (big Endian or little Endian); for word alignment (4-byte boundaries or 8-byte boundaries); for integer data-type length (16-bit, 32-bit, or 64-bit); and for doubles (byte-swapped or not).

By default, SAS creates a new SAS data set by using the data representation of the CPU that is running SAS. Specifying the OUTREP= option enables you to create a SAS data set with a different data representation. For example, in a UNIX environment, you can create a SAS data set that uses a Windows data representation. For more information about compatibility and data representation, see "Processing Data Using Cross-Environment Data Access (CEDA)" in Chapter 32 of *SAS Language Reference: Concepts*.

Values for OUTREP= are listed in the following table:

Table 2.1 *Data Representation Values for OUTREP= Option*

OUTREP= Value	Alias*	Environment
ALPHA_TRU64	ALPHA_OSF	Tru64 UNIX
ALPHA_VMS_32	ALPHA_VMS	OpenVMS Alpha
ALPHA_VMS_64		OpenVMS Alpha
HP_IA64	HP_ITANIUM	HP-UX for the Itanium Processor Family Architecture
HP_UX_32	HP_UX	HP-UX for PA-RISC
HP_UX_64		HP-UX for PA-RISC, 64-bit
INTEL_ABI		ABI for Intel architecture

OUTREP= Value	Alias*	Environment
LINUX_32	LINUX	Linux for Intel architecture
LINUX_IA64		Linux for Itanium-based systems
LINUX_X86_64		Linux for x64
MIPS_ABI		MIPS ABI
MVS_32	MVS	31-bit SAS on z/OS
MVS_64_BFP		64-bit SAS on z/OS
OS2		OS/2 on Intel
RS_6000_AIX_32	RS_6000_AIX	AIX
RS_6000_AIX_64		AIX
SOLARIS_32	SOLARIS	Solaris for SPARC
SOLARIS_64		Solaris for SPARC
SOLARIS_X86_64		Solaris for x64
VAX_VMS		OpenVMS VAX
VMS_IA64		OpenVMS on HP Integrity
WINDOWS_32	WINDOWS	32-bit SAS on Microsoft Windows
WINDOWS_64		64-bit SAS on Microsoft Windows (for both Itanium-based systems and x64)

* It is recommended that you use the current values. The aliases are available for compatibility only.

Details

CAUTION:

Transcoding could result in character data loss when encodings are incompatible. For information about encoding and transcoding, see *SAS National Language Support (NLS): Reference Guide.*

See Also

* Processing Data Using Cross-Environment Data Access (CEDA)
* "Definition of Cross-Environment Data Access (CEDA)" in Chapter 32 of *SAS Language Reference: Concepts*

POINTOBS= Data Set Option

Specifies whether SAS creates compressed data sets whose observations can be randomly accessed or sequentially accessed.

Valid in:	DATA step and PROC steps
Category:	Observation Control
Restriction:	POINTOBS= is effective only when creating a compressed data set. Otherwise it is ignored.

Syntax

POINTOBS=YES | NO

Syntax Description

YES

causes SAS software to produce a compressed data set that might be randomly accessed by observation number. This is the default.

Examples of accessing data directly by observation number are:

- the POINT= option of the MODIFY and SET statements in the DATA step

- going directly to a specific observation number with PROC FSEDIT.

    TIP Specifying POINTOBS=YES does not affect the efficiency of retrieving information from a data set. It does increase CPU usage by approximately 10% when creating a compressed data set and when updating or adding information to it.

NO

suppresses the ability to randomly access observations in a compressed data set by observation number.

TIP Specifying POINTOBS=NO is desirable for applications where the ability to point directly to an observation by number within a compressed data set is not important. If you do not need to access data by observation number, then you can improve performance by approximately 10% by specifying POINTOBS=NO:

- when creating a compressed data set

- when updating or adding observations to it

Details

Note that REUSE=YES takes precedence over POINTOBS=YES. For example:

```
data test(compress=yes pointobs=yes reuse=yes);
```

results in a data set that has POINTOBS=NO. Because POINTOBS=YES is the default when you use compression, REUSE=YES causes POINTOBS= to change to NO.

See Also

Data Set Options:

- "COMPRESS= Data Set Option" on page 13
- "REUSE= Data Set Option" on page 53

System Options:

- "COMPRESS= System Option" in *SAS System Options: Reference*
- "REUSE= System Option" in *SAS System Options: Reference*

PW= Data Set Option

Assigns a READ, WRITE, and ALTER password to a SAS file, and enables access to a password-protected SAS file.

Valid in:	DATA step and PROC steps
Category:	Data Set Control

Syntax

PW=*password*

Syntax Description

password
> must be a valid SAS name, which limits the password to eight characters and is case insensitive. See "Words in the SAS Language" in Chapter 3 of *SAS Language Reference: Concepts* .

Details

The PW= option applies to all types of SAS files except catalogs. You can use this option to assign a password to a SAS file or to access a password-protected SAS file.

When replacing a SAS data set that is protected by an ALTER password, the new data set inherits the ALTER password. To change the ALTER password for the new data set, use the MODIFY statement in the DATASETS procedure.

Operating Environment Information
> See the appropriate sections of the SAS documentation for your operating environment for more information about using passwords.

Note: A SAS password does not control access to a SAS file beyond the SAS system. You should use the operating system-supplied utilities and file-system security controls in order to control access to SAS files outside of SAS.

See Also

Data Set Options:

- "ALTER= Data Set Option" on page 8
- "ENCRYPT= Data Set Option" on page 17
- "READ= Data Set Option" on page 48
- "WRITE= Data Set Option" on page 68

Other:

- "File Protection" in Chapter 34 of *SAS Language Reference: Concepts*
- "Manipulating Passwords" in Chapter 15 of *Base SAS Procedures Guide*

PWREQ= Data Set Option

Specifies whether to display a dialog box to enter a SAS data set password.

Valid in:	DATA and PROC steps
Category:	Data Set Control

Syntax

PWREQ=YES | NO

Syntax Description

YES

    specifies to display a dialog box.

NO

    prevents a dialog box from displaying. If a missing or invalid password is entered, the data set is not opened and an error message is written to the SAS log.

Details

In an interactive SAS session, the PWREQ= option controls whether a dialog box displays after a user enters an incorrect or a missing password for a SAS data set that is password protected. PWREQ= applies to data sets with read, write, or alter passwords. PWREQ= is most useful in SCL applications.

See Also

Data Set Options:

- "ALTER= Data Set Option" on page 8
- "ENCRYPT= Data Set Option" on page 17
- "PW= Data Set Option" on page 47
- "READ= Data Set Option" on page 48
- "WRITE= Data Set Option" on page 68

READ= Data Set Option

Assigns a READ password to a SAS file that prevents users from reading the file, unless they enter the password.

Valid in:	DATA step and PROC steps
Category:	Data Set Control

Syntax

READ=*read-password*

Syntax Description

read-password
> must be a valid SAS name. See Rules for Words and Names in the SAS Language in *SAS Language Reference: Concepts*.

Details

The READ= option applies to all types of SAS files except catalogs. You can use this option to assign a password to a SAS file or to access a read-protected SAS file.

Note: A SAS password does not control access to a SAS file beyond the SAS system. You should use the operating system-supplied utilities and file-system security controls in order to control access to SAS files outside of SAS.

See Also

Data Set Options:

- "ALTER= Data Set Option" on page 8
- "ENCRYPT= Data Set Option" on page 17
- "PW= Data Set Option" on page 47
- "WRITE= Data Set Option" on page 68

Other:

- "File Protection" in Chapter 34 of *SAS Language Reference: Concepts*
- "Manipulating Passwords" in Chapter 15 of *Base SAS Procedures Guide*

RENAME= Data Set Option

Changes the name of a variable.

Valid in:	DATA step and PROC steps
Category:	Variable Control

Syntax

RENAME=(*old-name-1=new-name-1* < *...old-name-n=new-name-n*>)

Syntax Description

old-name
> the variable that you want to rename.

new-name
> the new name of the variable. It must be a valid SAS name.

Details

If you use the RENAME= data set option when you create a data set, the new variable name is included in the output data set. If you use RENAME= on an input data set, the new name is used in DATA step programming statements.

If you use RENAME= on an input data set that is used in a SAS procedure, SAS changes the name of the variable in that procedure. If you use RENAME= with WHERE processing such as a WHERE statement or a WHERE= data set option, the new name is applied before the data is processed. You must use the new name in the WHERE expression.

If you use RENAME= in the same DATA step with either the DROP= or the KEEP= data set option, the DROP= and the KEEP= data set options are applied before RENAME=. You must use the old name in the DROP= and KEEP= data set options. You cannot drop and rename the same variable in the same statement.

Note: The RENAME= data set option has an effect only on data sets that are opened in output mode.

Comparisons

- The RENAME= data set option differs from the RENAME statement in the following ways:

  - The RENAME= data set option can be used in PROC steps and the RENAME statement cannot.

  - The RENAME statement applies to all output data sets. If you want to rename different variables in different data sets, you must use the RENAME= data set option.

  - To rename variables before processing begins, you must use a RENAME= data set option on the input data set or data sets.

- Use the RENAME statement or the RENAME= data set option when program logic requires that you rename variables such as two input data sets that have variables with the same name. To rename variables as a file management task, use the DATASETS procedure.

Examples

Example 1: Renaming a Variable at Time of Output

This example uses RENAME= in the DATA statement to show that the variable is renamed at the time it is written to the output data set. The variable keeps its original name, X, during the DATA step processing:

```
data one;
   input x y z;
   datalines;
24 595 439
243 343 034
;
```

Example 2: Renaming a Variable at Time of Input

This example renames variable X to a variable named KEYS in the SET statement, which is a rename before DATA step processing. KEYS, not X, is the name to use for the variable for DATA step processing.

```
data three;
   set one(rename=(x=keys));
   z=keys+y;
run;
```

Example 3: Renaming a Variable for a SAS Procedure with WHERE Processing

This example renames variable Score1 to a variable named Score2 for the PRINT procedure. Because the new name is applied before the data is processed, the new name must be specified in the WHERE statement.

```
proc print data=test (rename=(score1=score2));
   where score2 gt 75;
run;
```

See Also

Data Set Options:

* "DROP= Data Set Option" on page 16
* "KEEP= Data Set Option" on page 30

Statement:

* "RENAME Statement" in *SAS Statements: Reference*

Procedure:

* Chapter 15, "DATASETS Procedure" in *Base SAS Procedures Guide*

REPEMPTY= Data Set Option

Specifies whether a new, empty data set can overwrite an existing SAS data set that has the same name.

Valid in:	DATA step and PROC steps
Category:	Data Set Control
Restriction:	Use with output data sets only.

Syntax

REPEMPTY=YES | NO

Syntax Description

YES

specifies that a new empty data set with a given name replaces an existing data set with the same name. This is the default.

Interaction: When REPEMPTY=YES and REPLACE=NO, then the data set is not replaced.

NO

specifies that a new empty data set with a given name does not replace an existing data set with the same name.

Tips:

Use REPEMPTY=NO to prevent the following syntax error from replacing the existing data set B with the new empty data set B that is created by mistake:

```
data mylib.a set b;
```

For both the convenience of replacing existing data sets with new ones that contain data and the protection of not overwriting existing data sets with new empty ones that are created by accident, set REPLACE=YES and REPEMPTY=NO.

Comparisons

- For an individual data set, the REPEMPTY= data set option overrides the REPEMPTY= option in the LIBNAME statement.

- The REPEMPTY= and REPLACE= data set options apply to both permanent and temporary SAS data sets. The REPLACE system option, however, only applies to permanent SAS data sets.

See Also

Data Set Options:

- "REPLACE= Data Set Option" on page 52

Statement Options:

- "REPEMPTY=YES|NO" in Chapter 2 of *SAS Statements: Reference*

System Options:

- "REPLACE System Option" in *SAS System Options: Reference*

REPLACE= Data Set Option

Specifies whether a new SAS data set that contains data can overwrite an existing data set that has the same name.

Valid in:	DATA step and PROC steps
Category:	Data Set Control
Restrictions:	Use with output data sets only.
	This option is valid only when creating a SAS data set.

Syntax

REPLACE=NO | YES

Syntax Description

NO

> specifies that a new data set with a given name does not replace an existing data set with the same name.

YES

> specifies that a new data set with a given name replaces an existing data set with the same name.

Details

* The REPLACE= data set option overrides the REPLACE system option for the individual data set.

* The REPLACE system option only applies to permanent SAS data sets.

Example

Using the REPLACE= data set option in this DATA statement prevents SAS from replacing a permanent SAS data set named ONE in a library referenced by MYLIB:

```
data mylib.one(replace=no);
```

SAS writes a message in the log that tells you that the file has not been replaced.

See Also

System Options:

* "REPLACE System Option" in *SAS System Options: Reference*

REUSE= Data Set Option

Specifies whether new observations can be written to freed space in compressed SAS data sets.

Valid in:	DATA step and PROC steps
Category:	Data Set Control
Restriction:	Use with output data sets only.

Syntax

REUSE=NO | YES

Syntax Description

NO

> does not track and reuse space in compressed data sets. New observations are appended to the existing data set. Specifying the NO argument results in less efficient data storage if you delete or update many observations in the SAS data set.

YES

> tracks and reuses space in compressed SAS data sets. New observations are inserted in the space that is freed when other observations are updated or deleted.

If you plan to use procedures that add observations to the end of SAS data sets (for example, the APPEND and FSEDIT procedures) with compressed data sets, use the REUSE=NO argument. REUSE=YES causes new observations to be added wherever there is space in the file, not necessarily at the end of the file.

Details

By default, new observations are appended to existing compressed data sets. If you want to track and reuse free space by deleting or updating other observations, use the REUSE= data set option when you create a compressed SAS data set.

REUSE= has meaning only when you are creating new data sets with the COMPRESS=YES data set option or system option. Using the REUSE= data set option when you are accessing an existing SAS data set has no effect.

Comparisons

The REUSE= data set option overrides the REUSE= system option.

REUSE=YES takes precedence over POINTOBS=YES. For example, the following statement results in a data set that has POINTOBS=NO:

```
data test(compress=yes pointobs=yes reuse=yes);
```

Because POINTOBS=YES is the default when you use compression, REUSE=YES causes POINTOBS= to change to NO.

See Also

Data Set Options:

- "COMPRESS= Data Set Option" on page 13

System Options:

- "REUSE= System Option" in *SAS System Options: Reference*

ROLE= Data Set Option

Identifies the fact table for a star schema join.

Valid in:	PROC SQL
Category:	Data Set Control
Restriction:	Use with input data sets only.

Syntax

ROLE= FACT | DIMENSION | DIM

Syntax Description

FACT
identifies the SAS data set as the fact table for a star schema.

DIMENSION | DIM
>   identifies the SAS data set as a dimension table for a star schema.

Details

A star schema is an arrangement of several tables in which a large fact table is joined to several dimension tables. For example, you can join SAS data sets by using SQL procedure syntax in order to create a star schema.

To improve the performance of the application that processes the joined tables, you can specify the ROLE= data set option. For example, you can specify ROLE=FACT to designate the specific fact table, or you can specify ROLE=DIMENSION to designate each dimension table, meaning that the table not designated is the fact table.

Because the role a table plays can change between queries, the ROLE= specification is in effect for the current step only and is not stored with the data set.

Example: Designating the Fact Table

In the following example, the ROLE= data set option improves the performance for PROC SQL. ORDERS is the fact table, and PRODUCT, PERIOD, and CUSTOMER are dimension tables.

```
proc sql;
   select orders.Order_Total
   from orders (role=fact), product, period, customer
   where orders.Product_ID = product.Product_ID
      and orders.Period_ID = period.Period_ID
      and orders.Customer_ID = customer.Customer_ID
      and product.Product_Name = "camera"
      and period.Period_Name = "1997"
      and customer.Customer_Name = "Walmart";
quit;
```

See Also

Chapter 7, "SQL Procedure" in *SAS SQL Procedure User's Guide*

SORTEDBY= Data Set Option

Specifies how a data set is currently sorted.

Valid in:	DATA step and PROC steps
Category:	Data Set Control

Syntax

SORTEDBY=_by-clause_ </ _collate-name_> | _NULL_

Syntax Description

by-clause < / *collate-name*>
>   indicates how the data is currently sorted.

by-clause
> names the variables and options that you use in a BY statement in a PROC SORT step.

collate-name
> names the collating sequence that is used for the sort. By default, the collating sequence is that of your operating environment. A slash (/) must precede the collating sequence.

> *Operating Environment Information*
>> For details about collating sequences, see the SAS documentation for your operating environment.

NULL
> removes any existing sort indicator.

Details

SAS determines whether a data set is already sorted by the key variable or variables in ascending order by checking the sort indicator. The sort indicator is stored in the data set descriptor information and is set from a previous sort. For detailed information about how the sort indicator is used and how it improves performance, see "The Sort Indicator" in Chapter 25 of *SAS Language Reference: Concepts* and "SORTVALIDATE System Option" in *SAS System Options: Reference* .

The following example of the CONTENTS procedure **Sort Information** section containing the **Validated** attribute set to NO, indicates that the data set was sorted using the SORTEDBY= data set option.

```
Sort Information
 Sortedby var1
 Validated NO
 Character Set ANSI
```

Comparisons

* Use the CONTENTS statement in the DATASETS procedure to see how a data set is sorted.

* The SORTEDBY= option indicates how the data is sorted, but does not cause a data set to be sorted.

Example

This example uses the SORTEDBY= data set option to specify how the data are currently sorted. The data set ORDERS is sorted by PRIORITY and by the descending values of INDATE. Once the data set is created, the sort indicator is stored with it. These statements create the data set ORDERS and record the sort indicator:

```
libname mylib 'SAS-library';
options yearcutoff=1920;
data mylib.orders(sortedby=priority
                  descending indate);
   input priority 1. +1 indate date7.
       +1 office $ code $;
   format indate date7.;
   datalines;
1 03may01 CH J8U
```

```
1 21mar01 LA M91
1 01dec00 FW L6R
1 27feb99 FW Q2A
2 15jan00 FW I9U
2 09jul99 CH P3Q
3 08apr99 CH H5T
3 31jan99 FW D2W
;
```

See Also

- Chapter 15, "DATASETS Procedure" in *Base SAS Procedures Guide*

- Chapter 48, "SORT Procedure" in *Base SAS Procedures Guide*

- Chapter 7, "SQL Procedure" in *SAS SQL Procedure User's Guide*

SPILL= Data Set Option

Specifies whether to create a spill file for non-sequential processing of a DATA step view.

Valid in:	DATA step and PROC steps
Category:	Data Set Control
Restriction:	Valid only for a DATA step view

Syntax

SPILL=YES | NO

Syntax Description

YES
creates a spill file for non-sequential processing of a DATA step view. This is the default.

Interaction: A spill file is never created for sequential processing of a DATA step view.

NO
does not create a spill file or reduces the size of a spill file.

Interaction: For direct (random) access, a spill file is always created even if you specify SPILL=NO.

Note: If you do not have enough disk space to accommodate a resulting spill file from a DATA step view that generates a large amount of data, specify SPILL=NO.

Tip: For SAS procedures that process BY-group data, consider specifying SPILL=NO in order to write only the current BY group to the spill file.

Details

When a DATA step view is opened for non-sequential processing, a spill file is created by default. The *spill file* contains the observations that are generated by a DATA step view. Subsequent requests for data read the observations from the spill file rather than execute the DATA step view again. The spill file is a temporary file in the WORK library.

Non-sequential processing includes the following access methods, which are supported by several SAS statements and procedures. How the SPILL= data set option operates with each of the access methods is described below:

random access

retrieves observations directly either by an observation number or by the value of one or more variables through an index without reading all observations sequentially. Whether SPILL=YES or SPILL=NO, a spill file is always created, because the processing time to restart a DATA step view for each observation would be costly.

BY-group access

uses a BY statement to process observations that are ordered, grouped, or indexed according to the values of one or more variables. SPILL=YES creates a spill file the size of all the data that is requested from the DATA step view. SPILL=NO writes only the current BY group to the spill file. The largest size of the spill file is a size to store the largest BY group.

two-pass access

performs multiple sequential passes through the data. With SPILL=NO, no spill file is created. Instead, after the first pass through the data, the DATA step view is restarted for each subsequent pass through the data. If small amounts of data are returned by the DATA step view for each restart, the processing time to restart the view might become significant.

Note: With SPILL=NO, subsequent passes through the data could result in generating different data. Some processing might require using a spill file; for example, results from using random functions and computing values that are based on the current time of day could affect the data.

Examples

Example 1: Using a Spill File for a Small Number of Large BY Groups

This example creates a DATA step view that generates a large amount of random data and uses the UNIVARIATE procedure with a BY statement. The example illustrates the effects of SPILL= with a small number of large BY groups.

With SPILL=YES, all observations that are requested from the DATA step view are written to the spill file. With SPILL=NO, only the observations that are in the current BY group are written to the spill file. The information messages that are produced by this example show that the size of the spill file is reduced with SPILL=NO. However, the time to truncate the spill file for each BY group might add to the overall processing time for the DATA step view.

```
options msglevel=i;
data vw_few_large / view=vw_few_large;
   drop i;
   do byval = 'Group A', 'Group B', 'Group C';
      do i = 1 to 500000;
         r = ranuni(4);
         output;
      end;
   end;
run;
proc univariate data=vw_few_large (spill=yes) noprint;
```

```
      var r;
      by byval;
   run;
   proc univariate data=vw_few_large (spill=no) noprint;
      var r;
      by byval;
   run;
```

Log 2.1 *SAS Log Output*

```
1     options msglevel=i;
2     data vw_few_large / view=vw_few_large;
3        drop i;
4
5        do byval = 'Group A', 'Group B', 'Group C';
6           do i = 1 to 500000;
7              r = ranuni(4);
8              output;
9           end;
10       end;
11    run;
NOTE: DATA STEP view saved on file WORK.VW_FEW_LARGE.
NOTE: A stored DATA STEP view cannot run under a different operating system.
NOTE: DATA statement used (Total process time):
      real time           21.57 seconds
      cpu time            1.31 seconds
12    proc univariate data=vw_few_large (spill=yes) noprint;
INFO: View WORK.VW_FEW_LARGE open mode: BY-group rewind.
13       var r;
14       by byval;
15    run;
INFO: View WORK.VW_FEW_LARGE opening spill file for output observations.
INFO: View WORK.VW_FEW_LARGE deleting spill file.  File size was 22506120 bytes.
NOTE: View WORK.VW_FEW_LARGE.VIEW used (Total process time):
      real time           40.68 seconds
      cpu time            12.71 seconds
NOTE: PROCEDURE UNIVARIATE used (Total process time):
      real time           57.63 seconds
      cpu time            13.12 seconds
16
17    proc univariate data=vw_few_large (spill=no) noprint;
INFO: View WORK.VW_FEW_LARGE open mode: BY-group rewind.
18       var r;
19       by byval;
20    run;
INFO: View WORK.VW_FEW_LARGE opening spill file for output observations.
INFO: View WORK.VW_FEW_LARGE truncating spill file.  File size was 7502040 bytes.
NOTE: The above message was for the following by-group:
      byval=Group A
INFO: View WORK.VW_FEW_LARGE truncating spill file.  File size was 7534800 bytes.
NOTE: The above message was for the following by-group:
      byval=Group B
INFO: View WORK.VW_FEW_LARGE truncating spill file.  File size was 7534800 bytes.
NOTE: The above message was for the following by-group:
      byval=Group C
INFO: View WORK.VW_FEW_LARGE deleting spill file.  File size was 32760 bytes.
NOTE: View WORK.VW_FEW_LARGE.VIEW used (Total process time):
      real time           11.03 seconds
      cpu time            10.95 seconds
NOTE: PROCEDURE UNIVARIATE used (Total process time):
      real time           11.04 seconds
      cpu time            10.96 seconds
```

Example 2: Using a Spill File for a Large Number of Small BY Groups

This example creates a DATA step view that generates a large amount of random data and uses the UNIVARIATE procedure with a BY statement. This example illustrates the effects of SPILL= with a large number of small BY groups.

With SPILL=YES, all observations that are requested from the DATA step view are written to the spill file. With SPILL=NO, only the observations that are in the current BY group are written to the spill file. The information messages that are produced by this example show that the size of the spill file is reduced with SPILL=NO. With small BY groups, this results in a large disk space savings.

```
options msglevel=i;
data vw_many_small / view=vw_many_small;
   drop i;
   do byval = 1 to 100000;
      do i = 1 to 5;
         r = ranuni(4);
         output;
      end;
   end;
run;
proc univariate data=vw_many_small (spill=yes) noprint;
   var r;
   by byval;
run;
proc univariate data=vw_many_small (spill=no) noprint;
   var r;
   by byval;
run;
```

Log 2.2 *SAS Log Output*

```
1     options msglevel=i;
2     data vw_many_small / view=vw_many_small;
3        drop i;
4
5        do byval = 1 to 100000;
6           do i = 1 to 5;
7              r = ranuni(4);
8              output;
9           end;
10       end;
11    run;
NOTE: DATA STEP view saved on file WORK.VW_MANY_SMALL.
NOTE: A stored DATA STEP view cannot run under a different operating system.
NOTE: DATA statement used (Total process time):
      real time           0.56 seconds
      cpu time            0.03 seconds
12    proc univariate data=vw_many_small (spill=yes) noprint;
INFO: View WORK.VW_MANY_SMALL open mode: BY-group rewind.
13       var r;
14       by byval;
15    run;
INFO: View WORK.VW_MANY_SMALL opening spill file for output observations.
INFO: View WORK.VW_MANY_SMALL deleting spill file.  File size was 8024240 bytes.
NOTE: View WORK.VW_MANY_SMALL.VIEW used (Total process time):
      real time           30.73 seconds
      cpu time            29.59 seconds
NOTE: PROCEDURE UNIVARIATE used (Total process time):
      real time           30.96 seconds
      cpu time            29.68 seconds
16
17    proc univariate data=vw_many_small (spill=no) noprint;
INFO: View WORK.VW_MANY_SMALL open mode: BY-group rewind.
18       var r;
19       by byval;
20    run;
INFO: View WORK.VW_MANY_SMALL opening spill file for output observations.
INFO: View WORK.VW_MANY_SMALL truncating spill file.  File size was 65504 bytes.
NOTE: The above message was for the following by-group:
      byval=410
INFO: View WORK.VW_MANY_SMALL truncating spill file.  File size was 65504 bytes.
NOTE: The above message was for the following by-group:
      byval=819
INFO: View WORK.VW_MANY_SMALL truncating spill file.  File size was 65504 bytes.
NOTE: The above message was for the following by-group:
      byval=1229

      .

    . Deleted many INFO and NOTE messages for BY groups

      .

INFO: View WORK.VW_MANY_SMALL truncating spill file.  File size was 65504 bytes.
NOTE: The above message was for the following by-group:
      byval=99894
INFO: View WORK.VW_MANY_SMALL deleting spill file.  File size was 32752 bytes.
NOTE: View WORK.VW_MANY_SMALL.VIEW used (Total process time):
      real time           29.43 seconds
      cpu time            28.81 seconds
NOTE: PROCEDURE UNIVARIATE used (Total process time):
      real time           29.43 seconds
      cpu time            28.81 seconds
```

Example 3: Using a Spill File with Two-Pass Access

This example creates a DATA step view that generates a large amount of random data and uses the TRANSPOSE procedure. The example illustrates the effects of SPILL= with a procedure that requires two-pass access processing.

When PROC TRANSPOSE processes a DATA step view, the procedure must make two passes through the observations that the view generates. The first pass counts the number of observations and the second pass performs the transposition. With SPILL=YES, a spill file is created during the first pass, and the second pass reads the observations from the spill file. With SPILL=NO, a spill file is not created—after the first pass, the DATA step view is restarted.

Note that for the first TRANSPOSE procedure, which does not include the SPILL= data set option, even though a spill file is used by default, the informative message about the open mode is not displayed. This action occurs to reduce the amount of messages in the SAS log for users who are not using the SPILL= data set option.

```
options msglevel=i;
data vw_transpose/view=vw_transpose;
   drop i j;
   array x[10000];
   do i = 1 to 10;
      do j = 1 to dim(x);
         x[j] = ranuni(4);
      end;
      output;
   end;
run;
proc transpose data=vw_transpose out=transposed;
run;
proc transpose data=vw_transpose(spill=yes) out=transposed;
run
proc transpose data=vw_transpose(spill=no) out=transposed;
run;
```

Log 2.3 *SAS Log Output*

```
1    options msglevel=i;
2    data vw_transpose/view=vw_transpose;
3      drop i j;
4      array x[10000];
5      do i = 1 to 10;
6        do j = 1 to dim(x);
7            x[j] = ranuni(4);
8        end;
9       output;
10     end;
11   run;
NOTE: DATA STEP view saved on file WORK.VW_TRANSPOSE.
NOTE: A stored DATA STEP view cannot run under a different operating system.
NOTE: DATA statement used (Total process time):
      real time            0.68 seconds
      cpu time             0.18 seconds
12   proc transpose data=vw_transpose out=transposed;
13   run;
INFO: View WORK.VW_TRANSPOSE opening spill file for output observations.
INFO: View WORK.VW_TRANSPOSE deleting spill file.  File size was 880000 bytes.
NOTE: View WORK.VW_TRANSPOSE.VIEW used (Total process time):
      real time            2.37 seconds
      cpu time             1.17 seconds
NOTE: There were 10 observations read from the data set WORK.VW_TRANSPOSE.
NOTE: The data set WORK.TRANSPOSED has 10000 observations and 11 variables.
NOTE: PROCEDURE TRANSPOSE used (Total process time):
      real time            4.17 seconds
      cpu time             1.51 seconds
14   proc transpose data=vw_transpose (spill=yes) out=transposed;
INFO: View WORK.VW_TRANSPOSE open mode: sequential.
15   run;
INFO: View WORK.VW_TRANSPOSE reopen mode: two-pass.
INFO: View WORK.VW_TRANSPOSE opening spill file for output observations.
INFO: View WORK.VW_TRANSPOSE deleting spill file.  File size was 880000 bytes.
NOTE: View WORK.VW_TRANSPOSE.VIEW used (Total process time):
      real time            0.95 seconds
      cpu time             0.92 seconds
NOTE: There were 10 observations read from the data set WORK.VW_TRANSPOSE.
NOTE: The data set WORK.TRANSPOSED has 10000 observations and 11 variables.
NOTE: PROCEDURE TRANSPOSE used (Total process time):
      real time            1.01 seconds
      cpu time             0.98 seconds
16   proc transpose data=vw_transpose (spill=no) out=transposed;
INFO: View WORK.VW_TRANSPOSE open mode: sequential.
17   run;
INFO: View WORK.VW_TRANSPOSE reopen mode: two-pass.
INFO: View WORK.VW_TRANSPOSE restarting for another pass through the data.
NOTE: View WORK.VW_TRANSPOSE.VIEW used (Total process time):
      real time            1.34 seconds
      cpu time             1.32 seconds
NOTE: The View WORK.VW_TRANSPOSE was restarted 1 times. The following view
statistics
      only apply to the last view restart.
NOTE: There were 10 observations read from the data set WORK.VW_TRANSPOSE.
NOTE: The data set WORK.TRANSPOSED has 10000 observations and 11 variables.
NOTE: PROCEDURE TRANSPOSE used (Total process time):
      real time            1.42 seconds
      cpu time             1.40 seconds
```

See Also

Data Set Options:

- "OBSBUF= Data Set Option" on page 32

TOBSNO= Data Set Option

Specifies the number of observations to send in a client/server transfer.

Valid in:	DATA step and PROC steps
Category:	Data Set Control
Restriction:	The TOBSNO= option is valid only for data sets that are accessed through a SAS server via the REMOTE engine.

Syntax

TOBSNO=n

Syntax Description

n

specifies the number of observations to be transmitted.

Details

If the TOBSNO= option is not specified, its value is calculated based on the observation length and the size of the server's transmission buffers, as specified by the PROC SERVER statement TBUFSIZE= option.

The TOBSNO= option is valid only for data sets that are accessed through a SAS server via the REMOTE engine. If this option is specified for a data set opened for update or accessed via another engine, it is ignored.

See Also

"FOPEN Function" in *SAS Functions and CALL Routines: Reference*

TYPE= Data Set Option

Specifies the data set type for a specially structured SAS data set.

Valid in:	DATA step and PROC steps
Category:	Data Set Control

Syntax

TYPE=$data$-set-$type$

Syntax Description

data-set-type

specifies the special type of the data set.

Details

Use the TYPE= data set option in a DATA step to create a special SAS data set in the proper format, or to identify the special type of the SAS data set in a procedure statement.

You can use the CONTENTS procedure to determine the type of a data set.

Most SAS data sets do not have a specified type. However, there are several specially structured SAS data sets that are used by some SAS/STAT procedures. These SAS data sets contain special variables and observations, and they are usually created by SAS statistical procedures. Because most of the special SAS data sets are used with SAS/STAT software, they are described in the *SAS/STAT User's Guide*. Some of the special data sets are CORR, COV, SSPC, EST, or FACTOR.

Other values are available in other SAS software products and are described in the appropriate documentation.

Note: If you use a DATA step with a SET statement to modify a special SAS data set, you must specify the TYPE= option in the DATA statement. The *data-set-type* is not automatically copied to the data set that is created.

See Also

- Chapter 12, "CONTENTS Procedure" in *Base SAS Procedures Guide*

WHERE= Data Set Option

Specifies specific conditions to use to select observations from a SAS data set.

Valid in:	DATA step and PROC steps
Category:	Observation Control
Restriction:	Cannot be used with the POINT= option in the SET and MODIFY statements.

Syntax

WHERE=(*where-expression-1<logical-operator where-expression-n>*)

Syntax Description

where-expression
is an arithmetic or logical expression that consists of a sequence of operators, operands, and SAS functions. An operand is a variable, a SAS function, or a constant. An operator is a symbol that requests a comparison, logical operation, or arithmetic calculation. The expression must be enclosed in parentheses.

logical-operator
can be AND, AND NOT, OR, or OR NOT.

Details

- Use the WHERE= data set option with an input data set to select observations that meet the condition specified in the WHERE expression before SAS brings them into the DATA or PROC step for processing. Selecting observations that meet the conditions of the WHERE expression is the first operation SAS performs in each iteration of the DATA step.

You can also select observations that are written to an output data set. In general, selecting observations at the point of input is more efficient than selecting them at the point of output. However, there are some cases when selecting observations at the point of input is not practical or not possible.

- You can apply OBS= and FIRSTOBS= processing to WHERE processing. For more information see "Processing a Segment of Data That Is Conditionally Selected" in Chapter 11 of *SAS Language Reference: Concepts* .

- You cannot use the WHERE= data set option with the POINT= option in the SET and MODIFY statements.

- If you use both the WHERE= data set option and the WHERE statement in the same DATA step, SAS ignores the WHERE statement for data sets with the WHERE= data set option. However, you can use the WHERE= data set option with the WHERE command in SAS/FSP software.

Note: Using indexed SAS data sets can improve performance significantly when you are using WHERE expressions to access a subset of the observations in a SAS data set. See "Understanding SAS Indexes" in Chapter 26 of *SAS Language Reference: Concepts* for a complete discussion of WHERE expression processing with indexed data sets and a list of guidelines to consider before indexing your SAS data sets.

Comparisons

- The WHERE statement applies to all input data sets, whereas the WHERE= data set option selects observations only from the data set for which it is specified.

- Do not confuse the purpose of the WHERE= data set option. The DROP= and KEEP= data set options select variables for processing, while the WHERE= data set option selects observations.

Examples

Example 1: Selecting Observations from an Input Data Set

This example uses the WHERE= data set option to subset the SALES data set as it is read into another data set:

```
data whizmo;
   set sales(where=(product='whizmo'));
run;
```

Example 2: Selecting Observations from an Output Data Set

This example uses the WHERE= data set option to subset the SALES output data set:

```
data whizmo(where=(product='whizmo'));
   set sales;
run;
```

See Also

- "WHERE Statement" in *SAS Statements: Reference*

- WHERE-Expression Processing

WHEREUP= Data Set Option

Specifies whether to evaluate new observations and modified observations against a WHERE expression.

Valid in:	DATA step and PROC steps
Category:	Observation Control

Syntax

WHEREUP=NO | YES

Syntax Description

NO

does not evaluate added observations and modified observations against a WHERE expression.

YES

evaluates added observations and modified observations against a WHERE expression.

Details

Specify WHEREUP=YES when you want any added observations or modified observations to match a specified WHERE expression.

Examples

Example 1: Accepting Updates That Do Not Match the WHERE Expression

This example shows how WHEREUP= permits observations to be updated and added even though the modified observation does not match the WHERE expression:

```
data a;
   x=1;
   output;
   x=2;
   output;
run;
data a;
   modify a(where=(x=1) whereup=no);
   x=3;
   replace; /* Update does not match WHERE expression */
   output; /* Add does not match WHERE expression */
run;
```

In this example, SAS updates the observation and adds the new observation to the data set.

Example 2: Rejecting Updates That Do Not Match the WHERE Expression

In this example, WHEREUP= does not permit observations to be updated or added when the update and the add do not match the WHERE expression:

```
data a;
   x=1;
   output;
   x=2;
   output;
run;
data a;
   modify a(where=(x=1) whereup=yes);
   x=3;
   replace; /* Update does not match WHERE expression */
   output; /* Add does not match WHERE expression */
run;
```

In this example, SAS does not update the observation nor does it add the new observation to the data set.

See Also

Data Set Option:

• "WHERE= Data Set Option" on page 65

WRITE= Data Set Option

Assigns a WRITE password to a SAS file that prevents users from writing to a file, unless they enter the password.

Valid in:	DATA step and PROC steps
Category:	Data Set Control

Syntax

WRITE=*write-password*

Syntax Description

write-password
> must be a valid SAS name.
>
> > **See:** Rules for Words and Names in the SAS Language in *SAS Language Reference: Concepts*

Details

The WRITE= option applies to all types of SAS files except catalogs. You can use this option to assign a password to a SAS file or to access a write-protected SAS file.

Note: A SAS password does not control access to a SAS file beyond the SAS system. You should use the operating system-supplied utilities and file-system security controls in order to control access to SAS files outside of SAS.

See Also

Data Set Options:

- "ALTER= Data Set Option" on page 8
- "ENCRYPT= Data Set Option" on page 17
- "PW= Data Set Option" on page 47
- "READ= Data Set Option" on page 48

Other:

- "File Protection" in Chapter 34 of *SAS Language Reference: Concepts*
- "Manipulating Passwords" in Chapter 15 of *Base SAS Procedures Guide*

Index

SAS® Publishing Delivers!

SAS Publishing provides you with a wide range of resources to help you develop your SAS software expertise.
Visit us online at **support.sas.com/bookstore**.

SAS® PRESS

SAS Press titles deliver expert advice from SAS® users worldwide. Written by experienced SAS professionals,
SAS Press books deliver real-world insights on a broad range of topics for all skill levels.

support.sas.com/saspress

SAS® DOCUMENTATION

We produce a full range of primary documentation:

- Online help built into the software
- Tutorials integrated into the product
- Reference documentation delivered in HTML and PDF formats—free on the Web
- Hard-copy books

support.sas.com/documentation

SAS® PUBLISHING NEWS

Subscribe to SAS Publishing News to receive up-to-date information via e-mail about all new SAS titles,
product news, special offers and promotions, and Web site features.

support.sas.com/spn

SOCIAL MEDIA: JOIN THE CONVERSATION!

Connect with SAS Publishing through social media. Visit our Web site for links to our pages on Facebook,
Twitter, and LinkedIn. Learn about our blogs, author podcasts, and RSS feeds, too.

support.sas.com/socialmedia

CPSIA information can be obtained at www.ICGtesting.com
Printed in the USA
LVOW05s0859090115

421944LV00002B/27/P